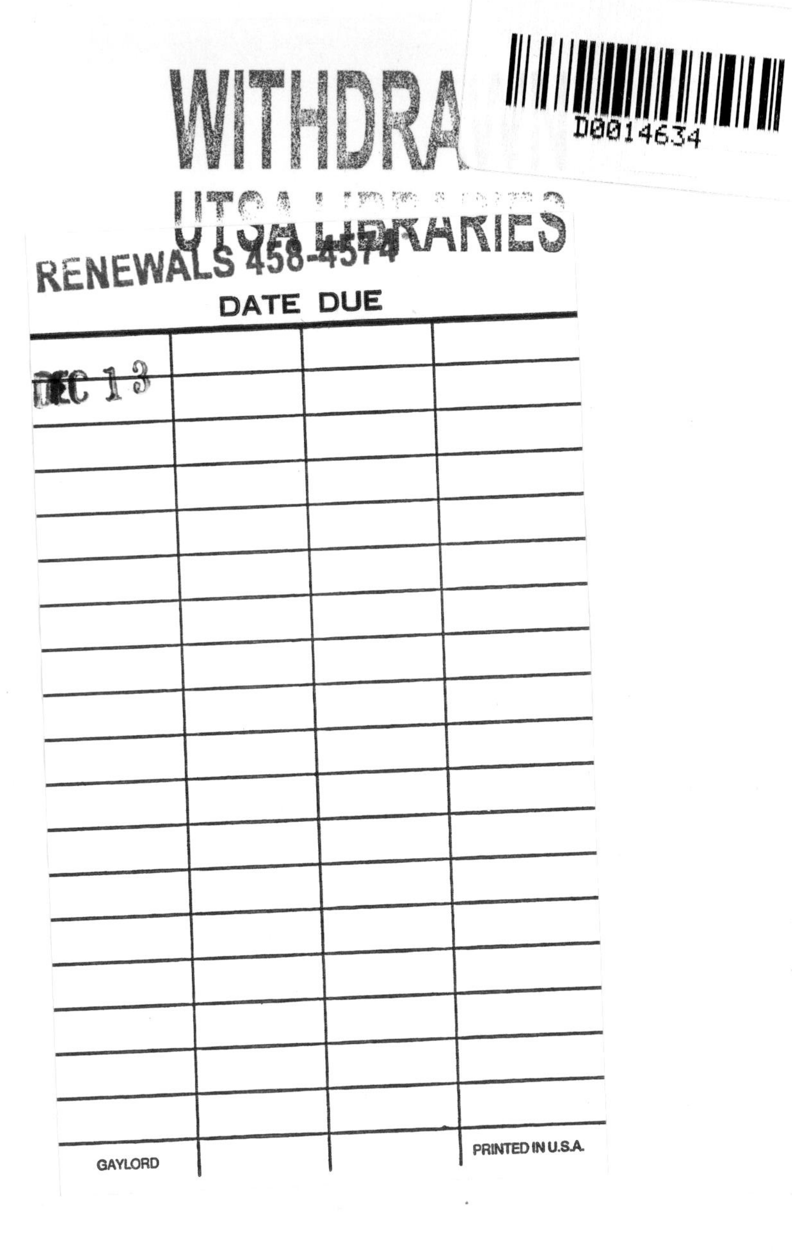

The Government of Science

The Government of Science

by Harvey Brooks

THE M.I.T. PRESS
Massachusetts Institute of Technology
Cambridge, Massachusetts, and London, England

Set in Linotype Electra
and printed and bound in the United States of America
by The Colonial Press Inc., Clinton, Massachusetts.

Library of Congress catalog card number: 68–22824

PREFACE

During the summer of 1967 I was approached by the M.I.T. Press with the suggestion that I prepare a collection of various essays and speeches in the area of science and public policy which I had written or delivered during the preceding seven years. This covers the period when I was closely associated with "science policy" in the executive branch of the federal government with the President's Science Advisory Committee, the National Science Board, and the Committee on Science and Public Policy of the National Academy of Sciences. Some of the essays represent adaptations from official documents or internal memoranda that I prepared during this period. Others are papers prepared on invitation for various conferences and symposia, which have previously been published in other collections. A few of the essays have borrowed quite freely from the ideas and even the happy turns of phrase provided for me by colleagues in correspondence and memoranda. While I take full responsibility for the way in which these ideas were woven into my own opinions and conclusions—often, perhaps, at variance with the original conception of their originators—I should like to take this occasion to acknowledge the debt. Help obtained from specific individuals is acknowledged in the introduction which precedes each essay.

The essays often reflect current preoccupations of federal science policy at the time they were written. They may thus to some extent appear dated. On the other hand, most of the problems dealt with are still with us, and likely to be so for some time, and as I reviewed the papers for publication, I found surprisingly little reason for altering either the opinions or the perspective in the light of recent developments. This may reflect only a hardening conservatism, or it may be that the principal themes are sufficiently persistent and pervasive to survive the twists and turns of contemporary political attention. Naturally I prefer the latter interpretation.

To the political scientist concerned with overall national policy, the choice of essays may appear to reflect excessive preoccupation with the problems of universities, particularly the sector of academic research. This preoccupation reflects my own interests and concerns as an adviser to government and as a university professor and administrator. Nevertheless, it also reflects my conviction that, despite the fact that it represents less than 10 percent of federal expenditures for science and technology, academic research has the most important consequences for the long-run health of our overall technical enterprise. In a remarkably short time the federal government has become the principal patron of science and of advanced education in the sciences and technology. This has happened with remarkably little conscious policy determination or debate. Yet it seems clear, to me at least, that we cannot retreat from this situation without disaster. For better or worse the "troubled partnership" between the major universities and the federal government is here to stay, and the central problem is how to maintain the momentum and the quality of the academic science enterprise through the period of continued growth that lies ahead.

Harvey Brooks

Cambridge, Massachusetts
January 29, 1968

CONTENTS

ONE □ The Government of Science

At the beginning of the Kennedy Administration in 1961 there was a rather searching review of the organization of the Executive Office for the coordination of national science policy. Various proposals for a Cabinet level Department of Science were seriously debated both within the Administration and within the Congress. The following chapter is a slightly edited version of a memorandum I prepared during the summer of 1961 for the President's Science Adviser, Dr. Jerome B. Wiesner, setting forth as objectively as I could the arguments both for and against a Department of Science. In reviewing this paper in the light of the experience of the past six years I find surprisingly little reason to alter the views expressed at that time. Some of the examples and some of the general intellectual and political climate toward science and technology now appear dated, but the basic conclusions and arguments do not seem to me to have been altered by subsequent events and experience.

INTRODUCTION

The phenomenal growth of the national scientific enterprise since 1950, especially that stimulated by federal support since 1957, has led to intensified discussion of the means by which this vast effort is planned and managed. Within the last few years, there has been a realization that while federal research and development expenditures represent a very modest fraction of national economic resources, they engage a much larger fraction of one of our scarcest national resources, namely, scientific and technical manpower. Furthermore, since the points of growth in our national economy appear to follow closely research and development expenditures, to the extent that these are channeled by decisions of the federal government, the whole thrust of our economy is determined. In sum, the social and economic leverage of the 2 percent of the gross national product which is expended on research and development by the federal government is out of proportion to the actual amount of money involved, yet the extent of this leverage is only now beginning to be appreciated.

Nevertheless, concern with our national scientific and technological strength, and with the influence of government upon it, has been manifest for some time. Many of the issues involved in the discussion of the management and planning of science in the federal government find a focus in the argument concerning whether there should be a Cabinet Department of Science. The present article is devoted to setting forth some of the pros and cons of such a department, not so much because I believe the issue itself is so central as because the arguments provide a framework within which it is easy to illuminate many of the problems and issues that are of current concern in the management of the federal science effort.

WHAT IS INCLUDED IN A DEPARTMENT OF SCIENCE?

Proposals for a Department of Science range all the way from very comprehensive centralization to relatively modest consolidation of a few of the more basically oriented government scientific activities.

There are currently four federal agencies whose mission is defined largely in terms of science: the Atomic Energy Commission, the National Aeronautics and Space Administration, the National Science Foundation, and the National Institutes of Health. Three of these are independent agencies reporting directly to the President, and the fourth is a part of a Cabinet Department. In addition to these major agencies, there are a number of scientific institutions, such as the National Bureau of Standards, which have a very broad capability and a present mission that is difficult to define in operationally useful terms within the framework of the department in which they are placed. In the most ambitious proposals for a Department of Science, the four agencies listed above are those usually mentioned for consolidation into a single Cabinet Department under a single Secretary. The various agencies in the Department of Defense are usually omitted from these considerations, despite the fact that this department was, prior to the spectacular growth of NASA, responsible for nearly 80 percent of federal expenditures for science and technology.

Indeed, the National Science Foundation, as originally envisioned in the report "Science the Endless Frontier," [1] had been expected to carry out specific research in support of health and defense missions, and it was only the long delay in the creation of NSF that resulted in the growth of independent basic research programs, first in the Navy, and later in the other military services and the Public Health Service.

In summary, a Department of Science would serve for the federal government a function analogous to that of the corporate research laboratory of a large private corporation, and the Secretary of Science would play a role analogous to that of the vice president for research of such a corporation. Creation of a Department of Science would not preclude operating departments from having their own separate laboratories rather strictly tied to the specific problems and missions of these departments, in analogy with the laboratories often associated with the manufacturing divisions of large corporations. Thus, for example, a laboratory like the Applied Physics Laboratory at Johns Hopkins or the Naval Ordnance Test Station at China Lake, California, would tend to remain an integral part of the Navy, while the Naval Research Laboratory, which is more in the nature of a corporate laboratory, would be transferred to the administration of a Department of Science.

Actually, there exists a whole spectrum of proposals of which that described in detail above is probably the most radical.

[1] Vannevar Bush, "Science the Endless Frontier, a Report to the President on a Program for Postwar Scientific Research," National Science Foundation (Washington, D.C., reprinted 1960).

A more modest proposal is for a Cabinet Department which would take over certain national laboratories having a rather broad capability, for example, the National Bureau of Standards, the Naval Research Laboratory, Lincoln Laboratory, the Brookhaven National Laboratory, and others, and would also take from existing agencies most of the contract research program in universities. Such a Cabinet Department might be similar to the National Science Foundation, enlarged to incorporate substantial inhouse capabilities and operating responsibilities over a broader spectrum of science, replacing the some seven agencies and offices which now play a significant role in the support of university science.

In almost any version of the Department of Science proposal, the new department would have responsibility for the present interagency scientific programs, such as oceanography, atmospheric sciences, high-energy nuclear physics, and so on, which are now coordinated through the Federal Council for Science and Technology. There would probably be fewer such programs because many present programs that now cut across agency lines would probably lie wholly within the assemblage of capabilities brought together under the direct management of a new department. In any case, the Department of Science would carry primary budgetary responsibility for interagency programs, and the funds for such programs would be defended by it before Congress and would be appropriated to it and allocated by it to the participating federal agencies. Through its reporting to Congress, it would take ultimate responsibility for the efficient management of such programs, and to that extent remove it from the agencies themselves.

Similarly, the Department of Science would take responsibility for certain government-wide activities in direct support of the national scientific enterprise, such as scientific information, recruitment of scientific manpower for the federal government, the support of scientific education, and so on.

ARGUMENTS FOR A DEPARTMENT OF SCIENCE

The following are some of the arguments that can be brought forward in favor of the creation of a Cabinet Department of Science:

1. It would ensure a better balanced national scientific program. With the present organizational arrangements, new and glamorous subjects, such as atomic energy and space, tend to be selected for special emphasis, often to the detriment of the balanced growth of basic science, and to the neglect of applied areas of equal or greater importance to national welfare. The accidents of congressional committee organization often tend to determine the relative allocation of resources among different fields of science without much reference to the real scientific opportunities or social needs involved. Scientific fields that can be made to appear to serve an immediately useful social or political goal receive lavish support while other fields of equal intellectual importance but less understandable to the public or to Congress receive only meager support. The generous support granted by Congress to the National Institutes of Health is contrasted with the very slow growth of the programs of the National Science Foundation, because the NIH programs are more understandable to the layman.

   In the present system there is often strong pressure to create a new agency for each new scientific discipline as its importance is recognized, and in this way to freeze into the executive branch a static organizational pattern which cannot accommodate itself readily to the dynamic reshuffling of relationships between fields which characterizes progress in science. Until relatively recently, many areas of applied science have been able to develop as somewhat isolated and self-contained disciplines without much dependence on the more fundamental sciences or on the general advance of science as a whole. Within the last twenty years this situation has entirely changed. Each applied area has drawn on a broader and broader base of fundamental science and reached further and further beyond empiricism and experience into common scientific principles. As a result of this, each new major governmental program places increasing demands on almost every branch of science and on advanced scientific education outside its own immediate domain. Whereas agricultural science, for example, was able to develop successfully as a self-contained specialty, "space science" really comprises almost every scientific and engineering discipline, both in the life sciences and the physical sciences. Thus, a government scientific agency can no longer control or command every technical capability or skill needed to carry out its assigned mission. The creation of new agencies for each new scientific discipline tends to place serious organizational barriers in the way of one agency's taking advantage of the skills and facilities of another. Furthermore, the United States has been noticeably slow in adopting and exploiting new areas of science or technology which do not fall clearly within the mission of an existing agency, for example oceanography, radio astronomy, and until the creation of NASA, the scientific exploration of space. It is often argued that a Department of Science could move much more rapidly into new areas and could continually reorganize itself to accommodate to the changing relationships between disciplines and the appearance of new disciplines.

2. A Department of Science would provide a more congenial home for certain national laboratories that cover a wide spectrum of disciplines. The National Bureau of Standards, the Brookhaven National Laboratory, and the Naval Research Laboratory could be cited as examples. In this connection a Department of Science would facili-

tate maximum national utilization of the full capabilities of these laboratories and would permit reassignment of laboratory missions to conform with the rapidly changing needs and requirements of technology. When a new national problem such as air traffic control, urban transportation, water or air pollution came to the fore, a Department of Science, it is argued, would permit us to move into the problem with all the national resources available, unconstrained by existing roles and missions. The potential contribution of a laboratory would be assessed wholly in terms of its capabilities rather than in terms of the limited mission of the agency of which it is a part.

3. Science budgets would be defended before Congress in a more uniform, coherent, and consistent way. There would be a single focus of responsibility in the executive branch, and this would engender greater congressional confidence in the overall management of the program and in the absence of "duplication and waste." There would be a single spokesman for science and technology in the executive branch, who could speak with the authority born of vast operational responsibility and budgetary control.

   Furthermore the creation of a single spokesman for science and technology in the executive branch would naturally lead to the development of a counterpart committee in Congress. There would thus grow up within Congress a group which would make a career of defending and promoting science as a whole and would provide a channel for mobilizing the testimony of the outside scientific community on congressional issues affecting the health of U.S. science. Much of this has already happened in the area of the health sciences in relation to the National Institutes of Health.

4. The centralization of key scientific service activities such as scientific information, the support of scientific educa-

tion, and the development of the overall scientific plant of the country would be greatly facilitated by a Department of Science and would provide greater insurance of the healthy development of science as a national resource. Such a department would pay greater attention to the health of scientific institutions.

5. A Department of Science could support and plan those technical activities which are of interest and importance to the government as a whole but not of overriding importance to any one agency or department. In this way it would be possible to avoid the difficult problem of adjusting agency interest and budgets to a comprehensive national program. Individual agencies could receive funds to support their role in an interagency science program outside their normal budget ceilings by direct transfer from the Department of Science. The problem of conflict of priorities would thus be avoided, and at the same time, the Department of Science could exercise much greater control over how the money was spent. The above function in the past has been exercised mainly through the Federal Council for Science and Technology, which has only the power of persuasion but no control over agency budgets. Coordination of interagency programs in the past has been successful only in those fields that were expanding very rapidly, because it is much easier to divide a pie which is growing in size 30 percent or 40 percent each year than to divide a pie of nearly constant size.
6. A single agency for science and technology would conserve scarce manpower needed in the effective monitoring and management of science programs in the federal government. Foı example, there are now no less than seven agencies that support substantial basic reseach programs in the physical sciences in universities: the Atomic Energy Commission, the Office of Naval Research, the Office of

Scientific Research in the Air Force, the Office of Army Research, the Advanced Research Projects Agency, the National Science Foundation, and the National Aeronautics and Space Administration. The administration of these extramural programs requires a high degree of skill, experience, and judgment, and the realization of the benefits of the basic research requires a unique combination of technical understanding, with knowledge of the needs of the government. As each new basic research agency has been created, it has recruited many of its key administrators from existing agencies with a resulting general dilution of talent and lowering of standards. It is argued that this talent should be concentrated in one agency where it can achieve maximum effectiveness.

It is also argued that the proliferation of agencies with different policies and administrative practices is demoralizing to the universities and greatly complicates their internal administrative problems.

In addition to these problems, there is also the problem that the basic research people in the more mission-oriented agencies are forced to spend a great deal of time and effort in defending basic research budgets against their superiors rather than on running the program. This happens because long-range research programs having a somewhat nebulous connection with specific mission requirements are forced to compete with urgent current problems and procurement in allocation of the budget. In a Department of Science as proposed, mission-oriented agencies would expend their basic research funds through the department and would thus not only make use of a single reservoir of administrative talent but also face a much stronger and more articulate set of defenders of the needs of basic research.

7. It is also argued that a new government agency is needed

now to assure the continued healthy growth of U.S. science. Since the last war, the spectacular growth of science has resulted mainly from the creation of a series of new scientific agencies at regular intervals. New money has been brought into the program by these new agencies rather than through expansion of the older agencies, which tend to reach a static budget after their glamour has worn off during the first few years. In each case the new agency has been created to exploit public interest in a new field or a new idea. The impact of all the series of new agencies has been to dramatize the importance of science to the public and to Congress. It is argued that the time is now ripe for the creation of still another new agency, and that because of general public acceptance of the importance of science, a Department of Science would enjoy support and backing which it could not have expected a few years ago.

ARGUMENTS AGAINST A DEPARTMENT OF SCIENCE

1. Science and technology are essentially tools for the achievement of social, political, or economic ends, whose desirability is arrived at through a political process.

   Essentially nonscientific ends are embodied in the missions of various government agencies which support scientific programs. It would be unhealthy and inefficient to deprive the mission-oriented agencies of one of the principal tools needed for accomplishing their mission. Even the agencies whose mission is defined mainly in scientific terms, such as the Atomic Energy Commission, have large operating and production responsibilities in addition to their research and development responsibility. It would be illogical and inefficient to attempt to separate these operating responsibilities from the research and development which support them, and yet unless these operating responsibilities are separated out, many of the arguments for a Department of Science lose much of their point.

The separation of research and development from operating missions would have one of two effects. Either the scientific effort financed within the Department of Science for the agencies would lose focus and purpose, and would thus become less effective in helping the agency to accomplish its mission, or else, more likely, the mission-oriented agency would "bootleg" its research in the guise of production or some other activity. Such bootleg research would be inefficient, done by the wrong kinds of people, and would lead to substantial duplication of the effort already going on in the Department of Science. We see evidence of this kind of duplication even now in connection with some of the major military and space hardware programs. It would be greatly aggravated and extended beyond the military sphere by the creation of a Department of Science.

2. Science and technology, regarded as ends in themselves, or as purely cultural activities, do not attract public support, at least on the scale which is now required. Support of science on this scale can only be sold to the public and to Congress by identifying it with specific desirable social goals such as the curing of disease, the enhancement of national security or national prestige, or the protection of public health or safety. We have seen many instances of this in the recent past. By identifying the solid-state sciences with the urgent practical materials needs of the Department of Defense, it was possible to achieve nearly a doubling of support of research in this area in some universities. The civilian nuclear power program of the Atomic Energy Commission has attracted wide public support because it was related to a simple and readily understood social goal. The program of the National In-

stitutes of Health has attracted congressional support much more readily than that of the National Science Foundation because it was easy to relate the work done, even the most basic work, to problems of health and disease, which were widely understood.

Some of the problems outlined above might be overcome by organizing the Department of Science in accordance with definite social objectives and goals rather than by scientific discipline. However, this type of organization might remove much of the advantage of flexibility which has been claimed for a Department of Science, and at the same time would not overcome the difficulty of the separation of operational from research and development functions.

3. Competition and diversity in the public support of science are important in ensuring its continued health and in the development of the most effective methods of administration and support. Historical experience suggests that conferring a functional monopoly on any agency in the federal government often leads to stagnation, inertia, and complacency. With the whole of American science now so heavily dependent on federal policies and programs, we cannot afford the risk of too much centralization of control, especially the risk of stagnation or political manipulation. Under the present system of basic research support by many federal agencies, individual agencies take great pride in the quality and productivity of the programs which they support and vie with each other in creating the conditions of administration which will attract proposals from the highest quality scientific groups. The inherent competitiveness of the scientific community has been matched by a healthy competitiveness within the government, which has led individual agencies to formulate their policies in such a way as to invite the confidence, approval, and praise

of the scientific community. Furthermore, the institutional and educational needs of science are quite diverse, and so the variation in policies which some complain about has certain advantages. Decentralized decision-making in the support of basic science certainly does create problems and results not only in some inefficiencies but also in undesirable effects on universities and research institutions. On the other hand, the decentralization of decision-making gives the scientific community a leverage on federal science policy which it would gradually lose were the policy centralized in a single agency. There is also an opposite danger that a Department of Science would become the captive of narrowly professional scientific concerns and interests and would cease to develop science in the best interests of the nation.

4. The imbalance between different scientific areas supposedly created by the present system of science support is probably not as serious in practice as it appears on paper. The missions of the Atomic Energy Commission, the Space Administration, and the Defense Department have provided a very broad stimulation to the physical sciences across the board, and there are few areas that have been seriously neglected as a result. Indeed, the glamourization of the missions of these agencies has probably resulted in more, rather than less, broad support for basic science. Many of the deficiencies noted in areas such as oceanography, geophysics, or atmospheric sciences have been due not so much to neglect as to the appearance of new opportunities opened up by massive progress in other areas of science or technology. Thus the appearance of such deficiencies should be regarded as a sign of the health of our whole scientific effort. If such deficiencies are recognized and met, little has been lost. As long as we maintain the quality of our whole scientific effort and training

at a sufficiently high level, we are in a position to make up newly identified deficiencies very rapidly, since well-trained scientists can channel their talents rapidly into entirely new areas.

The fashions in science, which often appear capricious to the layman, produce in practice a concentration of effort which leads to breakthroughs more rapidly and effectively than would a more centrally managed and less spontaneous effort. Scientific fashions and the rapid evolution and dissolution of communities of interest within science are strongly offsetting influences to the apparent high degree of institutional fragmentation in U.S. science, especially in the field of basic research.

5. The world scientific community constitutes an extremely complex social system, a subsystem within our whole society which is very little understood, least of all by scientists themselves. The present system of federal support of science has grown in an evolutionary way with relatively little conscious planning and has been the result of thousands of individual scientific and governmental decisions in response to immediately felt needs. Nobody is wise enough to foresee all of the effects of any organizational change at the federal level, especially when one factors in the unpredictable influence of individual personalities. It is more sensible for the government to make small organizational changes and arrangements in response to specific and clearly identified needs and deficiencies rather than attempt to mastermind or rationalize the whole process by setting up a radically new and apparently more logical organization whose effects would, in fact, be completely unpredictable. The creation of the Office of Special Assistant to the President, the President's Science Advisory Committee, the Federal Council for Science and Technology, and most recently, the Office of Science and

Technology are examples of evolutionary changes of the type that are most likely to meet the requirements for government planning for science. We need to create such institutions one at a time, and measure their influence on the scientific enterprise over a significant period. We also need to devise ways to make the most effective use of existing institutions.

Many deficiencies in our planning for science are the result of inadequate understanding of planning itself, of what things should be influenced by government and what things should be left to the natural responses of the scientific community. These deficiencies will not be removed by organizational changes but only by improved understanding of the relations between science and society.

It is possible that the present system for governing federal science is gradually evolving toward a Department of Science or something closely resembling it. If this is so, it will be much healthier if this evolution does not take place too rapidly or too radically.

6. The most serious management problems pertaining to government science and technology are related not to basic and applied research but rather to large development projects. The problems in this area are connected fundamentally with the choices among alternative goals rather than with specifically technical problems. Most of these choices involve economic evaluations (as in the case of civilian nuclear power) or operational cost-effectiveness studies (as in military and space systems). To an increasing degree these decisions depend as much on considerations of political, social, or military goals as on questions of technical feasibility. It is difficult to see how a Department of Science, which is further removed from these nonscientific aspects, could deal more effectively with this type of problem than the existing federal departments and

agencies. Indeed, one of the problems with which we are faced in the development of major systems is that technical feasibility tends to become confused with military or economic desirability. Technological developments tend to take on a life of their own, independent of the military, social, or economic context in which they will operate. The number of technical possibilities is rapidly exceeding the availability of resources to realize them, and more and more the problem of choice becomes a problem in resource allocation, an economic rather than a technical problem. The tendency for divorcement of technology from its political, social, or military context is likely to be aggravated rather than relieved by the creation of a Department of Science. There appears to be no good substitute for the present methods of debate and negotiation for resolving the complex interactions of technical and nontechnical considerations which are inevitably involved in all of our major decisions about priorities, whether between research fields, between hardware or operational systems, or even between research and procurement.

7. While the protection of the integrity of basic research is of the utmost importance, maintenance of a proper channel of communication from basic research to applications is also essential to the effective conduct of development. In the federal government, this channel is most effectively provided by the program officers who administer basic research for their mission-oriented agencies. It should be the duty of these program officers to understand the applied needs and requirements for their agency and to be alert to all the opportunities for filling these needs, which result not only from the basic research programs that they administer but also from related work throughout the whole body of science. It is their thorough knowledge of basic research and their contact with the

scientific community which give them the necessary communication with the scientific world to alert them to the opportunities provided by science, but they need also to understand enough of the mission of their agency to be able to match scientific opportunity to need. If all basic research programs were administered exclusively in the Department of Science, the vital channel of communication between basic and applied work would be weakened, since the program officers of the Department of Science, though highly competent in science, would not be thoroughly familiar with the needs and requirements of the various government agencies.

CONCLUSIONS

1. In the American system of government, central management of the scientific enterprise, even by scientists, cannot be an effective alternative to the complicated and often frustrating process of arriving at a national consensus. Science is an important instrument for almost all the goals of the federal government; the agencies responsible for the achievement of these goals cannot function effectively if they do not individually keep their channels of communication open to the world scientific community, which they can only do by carrying out or supporting research and development on their own.
2. Although the present diversity of support and decentralization of decision-making for science are desirable, further fractionalization of scientific support should probably be discouraged, and in general, new areas of science should be developed by existing agencies or by the interagency mechanism rather than by the creation of wholly new federal scientific agencies.
3. The creation of any new scientifically oriented federal

agency should be considered only when its service, production, or other operational functions reach an importance that is at least commensurate with its research and development function.

4. Better long-range planning for science and technology in the federal government is urgently needed, but in the last analysis, must be achieved by interagency agreement rather than by central direction. Many of the weaknesses noted in the present system for the management of science result from lack of technical competence or lack of adequate status for scientific activities within the agencies themselves rather than from deficiencies in central management and planning.
5. The function of central planning and coordination for science in the federal government is not to control the substance of the scientific activity in the nation but rather to ensure that the scientific enterprise as a whole develops in a way which is most responsive to the needs of the country and regulates itself responsibly. This function includes making sure that the needs and opportunities in science are made known and receive the proper attention in the process of arriving at a consensus on what the government should do. In the final analysis, continued and increasing support of science by the federal government will depend upon its continuing ability to demonstrate its social utility. Although the cultural and ethical aspects of science are of tremendous importance, one cannot expect that society will continue to support it on the present scale as a purely cultural activity. Therefore, in the management of science by the federal government attention must be given to the efficient utilization of science and to the realization of the opportunities it provides. Effective utilization does not automatically follow from a healthy and vigorous basic science, which is thus a necessary but not a sufficient condition.

TWO □ Science and the Allocation of Resources

The following paper, which appeared in the March 1967 number of the American Psychologist, *was originally prepared as a paper for oral delivery and discussion at a meeting of the* American Political Science Association *in September 1966. In many respects it deals with the same sets of themes and issues which were handled in the preceding chapter, "The Government of Science." The basic issue is the same, whether it is better to organize science and technology within the federal structure on the basis of technical themes or techniques or to organize around social purposes, with disciplines and techniques being subsidiary. This paper should be read as a companion to the earlier chapter.*

INTRODUCTION

Although science and technology have always played a prominent part in American government, the present system of government support for technical activities has grown up in a largely unplanned way as a series of responses to specific governmental needs as they arose and were perceived through political and administrative processes. There has never been any general theory of the relations of government and science which could be called a national science policy. Rather there was a series of science policies framed in the context of particular agencies in a manner largely incidental to their principal missions.

On the whole this spontaneous growth has been a source of vigor and strength. The dynamics of American science has not been inhibited by theoretical preconceptions as to its place in government or society, and room has been left for the maximum of individual initiative and experimentation.

Academic science and basic research grew up largely independent of the federal government, and the present structure of the American scientific establishment under federal support is essentially an inheritance from the earlier initiatives of private universities, the private foundations, and the state universities. Even though today federal funds support 70 percent of academic research and about 20 percent of university budgets, federal support has tended to strengthen the pre-existing pattern and structure without conscious consideration of the possibility of modifying the structure.

Thus, until World War II, federal policy toward science may be described as largely instrumental in character — that is, concerned with the utilization of science and technology for rather closely defined social purposes such as agriculture, defense, or natural resource development, which had received general political recognition as legitimate federal responsibilities. Federal technical activities were conducted largely within the structure of the Civil Service under close executive and congressional supervision. This situation was to some extent possible because of the nature of the applied sciences involved. It was possible to associate a fairly self-contained body of knowledge with a specific government mission, for example, agricultural science with agriculture, descriptive and historical geology with natural resources, certain specific mechanical arts with defense. Thus the applied sciences of concern to the federal government tended to be intellectually autonomous, with relatively little dependence on contemporary basic science or on each other. They were to a large extent empirical and descriptive, depending more upon the systematic accumulation and organization of data than upon the development of a highly articulated theoretical structure.

The idea of the government supporting research whose primary aim was the development of a theoretical structure which might be later applied was rather foreign to this system. Thus, although agricultural science did make important use of genetic principles and contributed toward their development, the support of genetics as such because of its importance to agriculture was not generally recognized as a federal responsibility. If one could gain understanding of genetics through the study of an economically important plant or animal species, this was all to the good. But it was not thought legitimate to support the study of genetics using an economically worthless organism, just because such an organism might be more suitable for achieving rapid advances in theoretical understanding.

Since in government it is natural to allocate resources in accordance with "missions" or social purposes, the problem of specific allocation of federal resources to science as such did not often arise as an identifiable political or administrative issue. The case for research was made within agencies in the context of their own missions, and the question of a common pool of knowledge which might serve many federal purposes simultaneously did not arise.

Of course, there is some oversimplification and exaggeration in the above picture. After years of debate Congress did accept the Smithsonian bequest, and the federal government thereby became the patron of science on its own terms. By the time of World War I there were federal laboratories such as the National Bureau of Standards which were doing important broad spectrum basic research. The Geological Survey had developed close ties with the academic community through the employment of faculty during the summer, and agricultural research was developing closer ties with the academic departments of biology in the land-grant universities. But, by and large, federal research was mission-oriented in the narrowest sense of the word, and each agency which used science at all tended to be scientifically self-sufficient. Relations with the outside scientific community were friendly, often cordial, but not close in any sense of mutual dependence.

WORLD WAR II AND BEYOND

The war years represented a remarkable watershed in American science policy, which was the result of a number of coinciding circumstances. In the first place the American academic scientific community in the late twenties and the decade of the thirties had begun to stand on its own feet intellectually in competition with Europe. Americans were beginning to be the trailblazers in nuclear physics, not only with the invention and perfection of new tools such as the cyclotron and the electrostatic accelerator but also with important contributions in pure theory. American work was in the lead in optical astronomy, largely owing to the construction of the big telescopes, and in medical research. This maturing basic research community was mobilized for war on an unprecedented scale, both in numbers and in degree of commitment. Partly this came about because the Hitler menace appeared uniquely directed at the traditions and ideals which seemed most important to the academic scientific community, both European and American, while the flow of prominent scientist refugees from Germany had given the threat an immediacy and reality which was a powerful stimulus to action. Science and technology proved more crucial to the immediate war effort than in any previous war; for the first time technical developments such as radar, the proximity fuse, and the atomic bomb went through their full cycle within the time span of the war and had a crucial effect on the military action. The scientists were involved much more deeply than merely solving problems posed by the military. They were involved directly in the solution of military problems and in the evolution of military tactics and strategy to take advantage of the new technical developments in weapons. The physics research of the 1930s had challenged electronic technology, and the experience gained by physicists in devising new instrumentation for their research proved uniquely useful in many of the new weapons developments. In a number of fields, most notably the case of the atomic bomb, science, which had been regarded as highly esoteric and remote from application before the war, was translated by forced draft into military hardware. The war saw the rapid evolution of an important new social invention, the research and development contract, which permitted great flexibility in the deployment of new skills for new purposes and in circumventing inherited organizational arrangements. For the first time in their experience scientists were provided with virtually unlimited resources, and saw what their skills could accomplish under these circumstances.

The American scientific community emerged from the war with a tremendous feeling of accomplishment and self-confidence — felt by the physicists most of all, but shared to some extent by all basic scientists. At this moment, before the relaxation of demobilization could be fully felt, the appearance of the cold war added impetus to the perpetuation of the arrangements and methods which had been developed during the war years. The policy thinking of physicists developed for the hot war tended to be perpetuated for the cold war, although the emphasis changed from weapons development to general-purpose science. There was widespread realization that the successful technical effort of the war years had drawn heavily on the bank of basic knowledge and research skills which had been built up during the fifteen years before the war, and a belief that some of the techniques which had been used for military research should be turned to replenishing this bank of basic knowledge. The freedom from minor bureaucratic interference which scientists had enjoyed under federal support during the war had broken down the previous resistance of the scientific and academic communities to this support, while at the same time the demonstration of what could be accomplished with large federal resources had made the return to a purely private basic research economy seem unattractive.

Thus the wartime experience of both federal administrators and scientists set the stage for the pattern of relations between the scientific establishment and the federal government which grew up during the subsequent twenty years. This pattern involved a continued strong dependence on the "R and D contract" as an instrument — with industry for development and applied research, and primarily with universities for basic research. It also involved a high degree of flexibility in the interpretation of mission relevance in the choice of what research might legitimately be supported by certain agencies, with scientists being accorded a great deal of freedom to develop their own approaches according to their own interests.

The change in pattern is, perhaps, most dramatically illustrated by the statistic that in 1938 about 40 percent of all federally supported research was provided by the Department of Agriculture, while by 1962, agricultural research constituted only 1.6 percent of federally supported research activity. In the late 1930s substantially all federally supported research was performed "inhouse." Even the NACA (National Advisory Committee on Aeronautics), which in many ways represented a foretaste of the postwar pattern of research, was about 98 percent intramural. By the 1960s only 14 percent of federally supported research was performed inhouse, and even if federally owned research centers operated under contract were included as inhouse, the figure did not rise above about 20 percent. Concomitant with the emphasis on extramural research, the intellectual direction of the effort fell much more heavily into the hands of the scientific community outside the federal establishment. A network of part-time advisory committees grew up in all the newer scientific agencies and eventually within the Executive Office. The National Academy of Sciences–National Research Council, a private agency operating under federal charter and with primarily federal support, grew to an $18 million a year science advisory agency, and a series of quasi-private advisory organizations like RAND or the Institute for Defense Analyses appeared on the scene. The system of selection of basic research proposals by juries of scientific peers grew up within the National Institutes of Health, and the National

Science Foundation extended a similar pattern of response to unsolicited research proposals from academic scientists in all fields. The whole structure expanded rapidly during the twenty-year postwar period, and such perturbations as shook it ended by accelerating its growth and reinforcing the general pattern of evolution which was already taking place. The biggest of these perturbations, the Sputnik panic of 1957, had its greatest ultimate effect on extramural research support, transformed NACA from an inhouse agency to an external contracting agency, brought nongovernment scientists into the White House, and generated the biggest single jump in the budget of the National Science Foundation. Although the Sputnik scare was largely the result of a failure in American technological planning, it was widely interpreted as a failure in education and basic research.

Since the middle sixties, a new wind has been felt across the scientific landscape. It is still difficult to foresee just what change in climate this wind will bring. But the relations of science and government and the nature of governmental planning for research and graduate education are being questioned with a skepticism and depth for which the scientific community was largely unprepared. For the first time there is a disposition to examine the federal R and D effort as a single coherent activity rather than as a separate series of activities, each related to a particular agency mission. Discussion has begun to center on comparison of the relative claims for support of different fields of science rather than, as in the past, the comparison between the scientific and operational activities of each agency. Although we have no Cabinet Department of Science, budgetary discussion has begun to be framed in terms of a collection of the scientific components of the budgets of all the agencies, thus producing the rudiments of a "science budget," and the concept of a "science budget" seems to be gaining ground all the time in thinking on the subject. There is much more political concern with the *institutions* of science, and such noncategorical issues as the geographical distribution of research funds, as contrasted with the substance and purpose of scientific activities supported by the federal government. In short, the scientific community and spokesmen in Washington find themselves faced increasingly with demands for the "planning" of science as such as opposed to the planning of the various federal activities to which science contributes. In the following section, we deal with some of the changes in the nature of modern science and technology, and of government organization for technical activities, which have contributed toward this identification of science as a separate and distinct function of government.

SOURCES OF PRESSURE FOR THE PLANNING OF SCIENCE

A primary source of the demand for more planning in science is the sheer magnitude of the federal government's technical activities and their rapid rate of growth. Although the growth of total R and D spending has leveled off during the last three years, research alone still continues to grow albeit at a slower rate, and the size of the technical community is growing faster than ever, though not faster than the "technical and professional" category of the labor force as a whole. The leveling off in federal spending appears to be attributable to the maturing of certain major space and defense systems, and when considered in the light of demands for new efforts in transportation, urban development, environmental pollution, weather modification, and exploitation of the oceans, may be only temporary, a pause for regrouping of R and D resources. In a booming economy, privately financed R and D is expanding as rapidly as ever in the past, and some private investment in technology is taking place in anticipation of future federal programs, most notably in oceanography and ocean engineering. The past statistics have been too often quoted to bear repetition here, but one should emphasize that while federal R and D represents about 15 percent of the administrative budget, it probably represents a much larger proportion of *discretionary* federal expenditures — that is, expenditures not completely committed by decisions in prior fiscal years.

A second important factor is that in no other major industry is the federal government such an exclusive customer. More than 50 percent of the nation's scientists and engineers are supported, directly or indirectly, from federal R and D expenditures, and large areas of induced or secondary demand in the private sector are produced by these expenditures; for example, in the scientific-instrument and electronic-component industries. There is increasing recognition that the nature of government investments in technology have a large impact on the direction and thrust of the private economy; indeed, have a much larger leverage on the growth and character of the economy than might be inferred from the relatively small proportion of the GNP which they represent. By and large many of the fastest growing industries are those which feed at least partially on government-induced demand, and the most successful competitors in the export market are often industries which depend strongly on government stimulated technology. Even if it is not clearly demonstrable, there is a widespread and growing belief that federal R and D expenditures have a large economic leverage in terms of regional growth, that there is a multiplier effect from R and D expenditures which does not result from other forms of federal investment. Considerations and beliefs such as these make it inevitable that Congress and the public are no longer content to see federal R and D expenditures purely in terms of their contribution toward the major goals of the agencies which make them. The economic and social leverage of such expenditures is regarded as too great to be left to the result of the interplay of relatively parochial agency interests, or the pressures of the constituencies which they serve.

A third factor creating a need for government-wide planning of technical activities is the increasing interdependence of the whole structure of science and technology. It is much less possible today than it was thirty years ago to associate specific areas of knowledge with specific federal missions. Individual agencies can no longer afford to be intellectually self-sufficient, either in the sense of dependence on other agencies or in the sense of dependence on nongovernment science and technology. Although today is popularly thought of, with some justification, as an age of increasing specialization, it is, paradoxically, also a day of disappearing barriers between scientific disciplines. In the life sciences, for example, there has been a revolution in the application of techniques and concepts of the physical sciences. Much of the instrumentation which is today used routinely in medical research, and even in diagnosis and care, was twenty years ago a homemade laboratory tool of physicists or chemists. Computer technology has touched all fields of knowledge, including the social sciences, but what has so far occurred is probably only the first faint glimmering of what is to come. The missions of many federal agencies reach more deeply into all areas of science and technology than in the past. The systems approach to many problems implies a much greater theoretical and analytical effort to explore feasibility and economy before embarking on the development and test of major systems. The feasibility exploration must often be based on a breadth and depth of understanding of modern science that would have been largely irrelevant in a simpler era, when actual construction and test of working models might have sufficed. The mere size, complexity, and cost of modern systems make a theoretical and analytical approach essential.

Equally dramatic as the application of physical techniques to the life sciences has been the revolution in methods of study of the environment. This is of special significance to federal programs because many of the areas in which a federal responsibility is most strongly recognized — such as pollution, urban development, weather prediction, or natural resource development — involve knowledge and understanding in the environmental sciences. These sciences have now outgrown an era which consisted largely of systematic description and data gathering and are entering a period where, with the aid of modern physical, chemical, and biological techniques and concepts, it is possible to develop a theoretical understanding of the environment as a total system. The oceans, the atmosphere, and the solid earth are interacting systems which cannot be considered in isolation from each other, and we are now beginning to realize that all are influenced by phenomena in nearby space and in the sun. The advent of high-speed computers, combined with new observational techniques such as satellites, constant density balloons, remote untended observation stations, and new types of sensors plus ultrareliable and miniaturized electronics, gives promise of a wholly new level of understanding and prediction of global atmospheric and oceanic circulation and the possibility both of really reliable numerical weather forecasting, and large-scale but safe experiments in weather and climate modification — safe because our understanding of atmospheric dynamics will be sufficiently advanced to permit assurance on a theoretical basis against undesirable side effects, or hitherto unpredictable consequences.

Thus to an increasing degree the progress of each given science or technology depends not only on its own internal development but on the progress of the whole intellectual structure of science. Not only does technology depend increasingly on contemporary science, but the advance of pure science depends increasingly on the application of new technologies. Agency missions depend on so many parts of science that no agency can afford to develop internally all the competences it needs to discharge its responsibilities effectively. Not only must it coordinate its technical activities with those of other agencies, but it must also maintain a close and continuing contact with scientists and scientific activities outside the government structure and outside the United States. While a great deal can be achieved through the normal processes of informal coordination and through the communications mechanisms of the world scientific community, it is inevitable that the interdependence of all areas of science and technology should generate a demand for planning of the federal technical enterprise as a whole.

One of the most significant responses of the federal structure to the increasing interdependence of technical activities and the increased intellectual structuring of science and technology has been the creation of new federal agencies defined by techniques or areas of science rather than social goals or missions. Perhaps the earliest example of such an agency was the NACA, created at the end of World War I to develop the technology of aviation on behalf of both military and civilian uses. This agency was, of course, transformed and adapted in 1958 to include the emerging technology of space, which in NASA has largely but not completely swallowed the aviation mission. The next agency was the Atomic Energy Commission, created to develop the whole science and technology of atomic energy and promote its use in as many applications as possible. Both atomic energy and space are *means* which serve many different social ends, and both agencies were assigned an explicit mission to promote and disseminate their particular technology and skills, and to find as many applications for them as possible, without restriction to any particular area of federal responsibility. Both agencies had a charter to, and did, enter many *functional* areas normally considered as belonging to the private sector or to other federal agencies. For example, the AEC took on responsibilities in the area of weapons development, nuclear propulsion for military purposes, civilian power, the use of isotopes for agricultural studies, the study of pollution by means of radioactive tracers, and studies in atmospheric diffusion. NASA entered the field of communications, of weather research, and of upper atmosphere exploration, overlapping with private industry, with the Weather Bureau, and with the Air Force. In all cases some effort was made to limit or contain the invasion of the missions of other agencies. In the AEC case joint offices were set up with other agencies in the areas of military propulsion, space propulsion, weapons development, and water desalting. By legislation the AEC was given a mission to transfer reactor technology as rapidly as possible to private industry. Similarly, as soon as NASA had brought space technology to the point where civilian communication by satellite looked feasible, a new quasi-private agency was created to exploit and market the resultant technology. Responsibility for the operation of a working weather satellite system was transferred to the Weather Bureau. But the basic budgetary allocations to AEC and NASA tended inevitably to be made in terms of means rather than ends.

There is now considerable pressure from many directions to repeat the pattern of AEC and NASA for other specific technologies. This has the longest history in the case of oceanography. Traditionally the marine sciences have served the missions of a large number of federal agencies, some twenty-two being involved in some aspect of the support of oceanography. Beginning with the upsurge of interest in oceanography following the National Academy report in 1958, an attempt was made to follow the traditional method of government planning by keeping the appropriate parts of oceanography within the agencies whose mission they served, and by attempting to superpose a coordinating group over the whole activity. This coordinating group was a forum for exchange of information and plans and for adjudication of conflicting or overlapping responsibilities, but it had no executive power other than the intangible prestige of location in the Executive Office. It had no budget of its own, and no clear responsibility for defending its decisions or programs before Congress. Oceanography within each agency had to be defended in terms of the total responsibility of that agency in relation to its budgetary ceiling rather than in relation to other parts of oceanography or in comparison with other fields of science or technology. Only within NSF was it really possible to make an effective comparison between oceanography and other fields of science.

Two arguments are made for the creation of separate agencies to deal with specific technologies across many social purposes. The first is that a new and emerging technology does not receive adequate attention unless the existing capabilities are collected in a single agency. The applications and ramifications of a new technology are not readily foreseeable, and it is consequently difficult to make a proper allocation of support between the many and diverse potential users of a new technology. Existing agencies are wedded to their own traditional techniques and are likely to resist new methods until they have been brought to a certain stage of demonstration. A new technology is a hothouse plant which has to be cultivated under glass, as it were, until it is healthy enough to be transplanted into the competitive environment of existing mission-oriented agencies. This general line of reasoning is well exemplified in the history of AEC. It proved to be one of the most significant and practical arguments for a separate agency under civilian control. A second related argument is more applicable to NASA. This is the case where a new technology is so complex and expensive, and yet useful to so many missions, that one cannot afford to develop it separately for each mission. The alternative of assigning responsibility to a lead agency — to the Air Force in the case of space, for example — is deemed undesirable because such an agency has its own mission-related interests and is unlikely to realize the full range of potential applications for a new technology unless it has responsibility for development and promotion of the technology itself without reference to any one mission. The commonality which unifies activity in a single agency is a set of techniques and skills rather than a defined set of goals. In this connection it is interesting to note the suggestion of the recent report of the President's Science Advisory Committee on oceanography[1] that the commonality of methods and techniques in the environmental sciences as a whole justifies the assemblage of all these sciences — including oceanography, atmospheric sciences, geology, marine biology, and upper atmospheric physics — into a single environmental science agency charged with the development of all these interacting sciences as a coherent discipline and with promoting their application to many federal missions. The proposal leaves the interface with existing activities somewhat vague, but implies continuation of such existing activities when sufficiently clearly related to presently defined missions. Thus the new agency would be concerned with generating new missions from a developing body of technology, while the older agencies would be concerned with utilizing and developing the environmental sciences for their particular existing missions. Whether this is a viable distinction in practice, of course, is the crux of the debate which will undoubtedly be stimulated by this proposal.

The principal argument against the creation of new agencies for new technologies is that they tend to take on a life of their own, the promotion of the technology becoming an end in itself apart from the social goals which it serves. What starts as a hothouse plant often grows into a rampaging weed. The cultivation of a technology, which is a secondary goal, becomes a primary goal. There is no place, no strong group, to which the

[1] Panel on Oceanography, President's Science Advisory Committee, "Effective Use of the Sea," The White House, June 1966 (U.S. Government Printing Office, 1966), pp. 88–90.

President or Congress can turn for unbiased and authoritative advice concerning the relative importance of different means to the same ends. For example, nuclear energy is only one of many methods for achieving economic central station power, and yet within the government it is difficult to find a technically strong group which can provide disinterested judgments on whether it is more important to push nuclear energy or new coal technologies, or whether innovation in transmission methods is more important than innovation in energy sources. The advice of the AEC must be regarded as suspect in this regard as long as its major mission is the promotion of nuclear energy rather than the development of economic power sources. Since independent agencies concerned with means or techniques also imply separate budgeting for such means, the particular techniques which are the province of separate agencies enjoy a preferential position for use by other mission-oriented agencies. If the cost of nuclear weapons is borne by the AEC, at least in part, then nuclear weapons enjoy a preferential position in the evaluation of weapons systems for Defense, since in the context of the military service budgets they come "for free." Such difficulties are not insurmountable, and are now being offset by such devices as program budgeting, but it is clear that when there are large agencies organized around means rather than ends, strong central planning at a higher level is necessary to ensure that the competing means are kept in proper relation to each other within the framework of national goals. Indeed, this is probably one of the most important sources of pressure for centralized planning of science, and if present public pressures for other independent technologically oriented agencies are successful, the requirement for a stronger planning mechanism will become even more urgent.

So far most of the pressures for planning I have discussed relate to applied science and technology. The "means" agencies referred to deal with a vertically integrated system ranging from nearly pure science to operations. What about planning for "pure" science? As long as basic science was largely in the hands of individual scientists working independently with a few colleagues or students, it required little planning or coordination. The built-in sociology and ethics of basic science tend to militate against duplication and to put a premium on every scientist remaining fully aware of all the work relevant to his own research. There is little wasted effort other than that inherent in a highly speculative and unpredictable activity. But the past fifteen years have seen even pure science carried out on an entirely new scale. We have the new phenomenon loosely known as "big science," that is, pure science carried on with complex and expensive equipment, and with a large supporting technological effort. In order of cost, the most important examples of such big science are space science, high-energy physics, oceanography, radio astronomy, and optical astronomy. These activities require planning for several reasons. In the first place, the design and development of complex equipment and the assembling of new research institutions to operate it require lead times of up to six or seven years, and the time is continually lengthening. The present proposed 200 Bev. accelerator would not become fully operative in a scientific sense until nearly 1975. In the second place, the equipment is usually so expensive that it often makes sense to construct only one of a kind. Even if expense were not a limitation, the variety of really significant experiments which can be done may not justify duplication of the same equipment from a strictly scientific viewpoint. In the third place, commitment to the initial capital investment implies a continuing commitment for operations which can range from one fifth to one half the capital cost. In addition to straight operating costs there is usually a continuing requirement for updating and for new types of auxiliary instrumentation. Each such investment thus entails a built-in escalation of the operating budget for pure science which will compete with alternate scientific activities long into the future if funds are limited. The problem of planning is compounded if different activities in big science are in different agencies, since they must compete both with each other across agency lines and with the other activities of a given agency. Thus high-energy physics inevitably competes with, say, nuclear rocket development or civilian reactors, but it also competes, in some sense at least, with space science or radio astronomy, in NASA and NSF, respectively. Such activities are sufficiently discrete and visible so that they cannot, as in the case of "little science," be treated as "technical overhead" on much larger applied and operational programs. Thus the growth of big science is a fourth factor leading toward greater centralized planning for science. It must be remembered also that the highly visible big-science activities are only the most obvious manifestation of an escalation in the scale of effort in pure science which is taking place all along the line. In almost every field of activity the individual scientist in his laboratory requires a much bigger capital investment in equipment and instrumentation than was the case a few years ago, and the growth of "team research" is everywhere evident. At any stage of its development each science is characterized by a "critical size," a minimum scale of effort, which is necessary for progress. To some extent, of course, this is a competitive phenomenon. Science is competitive as well as cooperative, and if one group achieves a certain scale of effort and instrumentation, others will have to follow if they hope to remain at the forefront of a field.

A fifth factor leading toward more planning is the entrance of the federal government on a large scale into the support of graduate education in science. Early support of university research was highly selective and essentially an increment added to accelerate an already existing activity. University research was supported to accelerate the achievement of certain research results regarded as important to mission-oriented agencies. But as the support of academic research grew at the rate of 20 to 25 percent a year for many years, the federal contribution became the main guarantor of the quality of graduate education in science and engineering. Though only a relatively modest fraction of the total cost of graduate training, it provided much of its thrust and direction, probably having a leverage well beyond its nominal financial impact. Thus what started as assistance of a special purpose and peripheral nature became institutionalized in the whole university system, but without conscious attention to institutional development as such. An extraordinary development of both the quality and quantity of American graduate education was the result, but it came as the largely unplanned by-product of funds spent for a quite different purpose. It was only a question of time before this process would begin to be viewed more self-consciously and in a more political context, and this is what is now happening. Thus the almost absentminded intervention of the federal government in higher education has created pressures for more "planning" of this process in an institutional and political sense.

A sixth factor for planning in science has been the emergence of international programs and the use of scientific cooperation as a tool of international relations. Such programs involve advance commitments on the part of the United States, often of several years' duration. They require complex logistic support, and carefully timed and coordinated field observations. Typically, most such programs have been in the physical and biological field sciences, because this is the area in which internationally coordinated observations have the biggest scientific payoff. Because such programs require long-range advance commitments for support, they tend to acquire a *de facto* priority which may cause budgetary problems with other less visible branches of science. But in any case international programs do not conform to the classic concept of science programs which compete with other activities directed toward a mission within the context of a single agency.

Finally, the most recent development creating pressure for more centralized planning is the emergence into public attention of a series of major social problems for which technical inputs are essential — environmental pollution, urban transportation, water resource development, poverty, regional economic development, rationalization of medical care. Many of these problems have been with us for a long time, but recent successes with the "systems" approach to weapons development and manned space exploration have stimulated hopes that a similar "forced development" approach can be applied to some of the troublesome problems in the civilian and public sector. If $5 billion can get a man on the moon by 1970, it is asked, why cannot a similar well-directed effort eradicate poverty or make our cities more habitable? The validity of the parallel between space exploration and urban development may be questionable, but the fact remains that it is drawn in the public mind, and the application of science to these problems has a political appeal which provides a forceful impetus toward centralized planning. While it is true that the achievement of such goals requires a good deal besides research — indeed, in the minds of many, research is the least difficult part of the problem — the feeling has arisen that somehow the inadequacies of the federal government in dealing with these problems stem, at least in part, from a misallocation of technical resources.

This feeling appears also to be reflected in recent legislation extending the authorization of the NSF to operate in the field of applied science. Although in its final form this authorization is restricted primarily to academic research, the legislative history suggests that some of the witnesses and some of the congressmen envisioned a considerably larger role. Implicitly the argument for this involves the assumption that agencies with social goals cannot be relied upon to generate all the research effort and ideas they need to accomplish their missions. The supposed backwardness of the Post Office in utilizing modern technology is often cited as an example. Reliance solely on mission-oriented agencies implies that research goals are *deducible* from social goals. While in many cases this may be true, it is also often true that research turns up new goals or modifies old ones. In other words, the definition of goals is itself partly the result of research. Presumably, the type of research to be done under NSF auspices would be that needed to define goals. This is achieved by basic as well as applied research. A good illustration of the situation may be provided by the historical role of NSF in the field of weather modification. NSF was assigned this responsibility at a time when it was not clear whether weather modification was feasible, and even if feasible, what particular types of objectives were most worthwhile. The time certainly was not ripe for operational experiments, and it was not even clear that a sensible goal of weather modification existed. NSF served as an agent for exploration of the field without final commitment to any particular technical or social goals. Thus the engagement of NSF in applied research in other areas might serve a dual purpose. On the one hand, it can develop a technology in the early stages before problems are clearly identified or goals defined — essentially the predefinition phase of applied research. On the other hand, it can avoid premature commitment to technological or social goals which are not yet demonstrably feasible. Nevertheless, it is also clear that the prospect of the entry of NSF into applied research poses new problems for centralized planning, since inevitably its activities would tend to invade the responsibilities of mission-oriented agencies, and problems of interfaces with such agencies become increasingly difficult and complex.

THE EXISTING SYSTEM OF DECISION-MAKING FOR SCIENCE

The alternatives before us have been succinctly summarized by William Capron[2] in testimony before the Reuss committee:

There seem to me to be two basically different ways in which the Executive and the Congress can go about the task of deciding how much to spend in any year on specific research and development undertakings: The first of these, and the way we handle this problem under present organizational and budgetary procedures, is to consider R. and D. as an aspect of each Federal activity: we consider the role of R. and D. in relation to the achievement of the objectives of each major program, e.g. Defense, Agriculture, etc. We make judgments, at least implicitly, about the relative importance or priority of various Federal programs and therefore about how to allocate tax dollars by considering *total* outlays (including R. and D.) needed to achieve program purposes. We may compare, say, expenditures on highways with expenditures on education. But the present approach to decision making does *not* lead us to focus directly on highway *research* outlays in relation to those for education research.

The other alternative, which Capron ascribes to Congressman Reuss, "seems to suggest an *R. and D. Budget* which would bring together Federal R. and D. support proposed for each agency, and . . . have the Executive Branch and the Congress make trade-offs not between *total* program spending, but directly between the R. and D. components of such spending."

Capron then goes on to make the important point that "one cannot make any judgment at all about the adequacy of research expenditures in a given field simply by looking at society's evaluation of the importance of the activity . . . it is not only the importance of a national goal but *the prospect of significant improvement in our ability to achieve that goal* as a result of R. and D. which determines whether a given level of R. and D.

[2] Capron, W. M., Statement of William M. Capron before a Subcommittee on Government Operations of the House of Representatives, January 7, 10, 11, 1966.

effort is proper either absolutely or in relation to R. and D. devoted to other purposes."

I believe this is a good statement of issues involved in the allocation of R and D resources, but, as indicated by the discussion in the preceding section, the simple dichotomy posed by Capron is no longer entirely realistic. The existence of separate budgetary units devoted to secondary goals, that is, to means rather than ends, makes the kind of trade-off suggested by Capron somewhat artificial. The entire program of NASA consists of R and D by definition. Even if one were to accept the objective of landing a man on the moon as a social purpose analogous to defense or agriculture, the scope of the NASA program extends considerably beyond this simple objective. The dilemma is well exemplified by the current debate over how the nation should utilize the capabilities developed in the Apollo program once the immediate objective is attained. As the program evolves, the NASA effort looks more and more like a solution looking for a problem. Similarly, as the mission-oriented agencies become more deeply involved in higher education, it is increasingly difficult to evaluate their research programs purely in relation to their operational activities without consideration of their impact on educational institutions. The mere existence of a large federal program, including that of NSF, which is science-oriented rather than mission-oriented, precludes the simple type of trade-off which Capron prefers.

Weinberg[3] has developed an ingenious concept for dealing with this difficulty, based on the idea of "technical overhead." He regards science not only as a direct cost attributable to national goals, in the sense implied by Capron, but also partly as an overhead cost, allocable but not attributable, to any one goal. That part of science, mostly basic science, for which the relationship between particular scientific activities and par-

[3] Weinberg, A. M., "Science, Choice, and Human Values," *Bulletin of the Atomic Scientists*, 22, 1966, pp. 8–13.

ticular practical goals is subject to a high degree of uncertainty is thus treated as an overhead rather than as a direct cost of the mission. In this concept even the purest science is to be viewed as "a general overhead to be assessed against the entire scientific technical enterprise because it supplies style and sophistication and standards." The basic idea behind the Weinberg thesis is that each successive level of management in the total scientific enterprise has a broader view of the whole, and will thus tend to add its own increment of general purpose basic (and often also applied) research in order to ensure a wide enough range of future options for the achievement of its goals. Weinberg's idea gets around some of the philosophical objections to the simple form of Capron's thesis, while still maintaining the present basic pattern of allocation. However, Weinberg's theory is not very helpful in a quantitative sense. The concept of science as "overhead" implies allocation on the basis of a percentage of some underlying activity, but it gives little basis for arriving at what this percentage should be, or how the overhead should be allocated to various types of institutions. Weinberg's proposed criteria of choice between fields do give some basis for allocating the overhead on the basis of subject matter.

One may also say that Weinberg's overhead idea is not so much a prescription for policy as it is a rationalization of existing policy. It provides the rationale by which the support of basic and academic research can be fitted in practice into the Capron scheme. Weinberg's explanation of the decision-making process also explains some of the administrative difficulties which beset the role of NSF. In the hierarchy of levels of management, each of which adds its increment of basic research, NSF should be above the mission-oriented agencies which support science, since it is supposed to have a broader view of total national needs and thus to add the increments of basic research which are not viewed as sufficiently relevant by individual agencies. In practice, however, NSF is one agency among equals, and with a smaller budget and less leverage than most. As an agency it has neither the power nor the government-wide perspective to exercise the general overhead function without strong and preferential support from the Executive Office of the President and indeed also from Congress. Only the Executive Office has the global perspective envisioned by Weinberg, but no discretionary funds of its own. Weinberg's theory thus explains why presidential intervention has recently played a crucial role at several points in the determination of the NSF budget, not only with respect to the total, but also with respect to the initiation and support of new activities such as the International Geophysical Year or the International Indian Ocean Expedition. There is a serious question, however, whether the NSF role among science-supporting agencies can long survive without its special relation to the Executive Office, even though nominally on a par with other agencies. Unfortunately, there is not a corresponding special relationship with Congress, which will be ultimately needed to stabilize its role in the allocation process.

What we seem to have at present is a "mixed" system of decision-making for science, which follows neither the Capron formula nor the formula ascribed to Reuss. In the budgetary system there is a strong intercomparison between agencies, but superposed upon this a weak intercomparison within parts of the R and D program. Consideration of the trade-offs *within* R and D seldom produces more than minor perturbations of the budget arrived at through the competition between agencies. These minor perturbations should not be minimized, however, for although they may have little influence in any one budgetary year, the cumulative impact over many budget years can be quite significant. The trade-offs within R and D tend to be determined at the Executive Office level, particularly by the interaction of the Office of Science and Technology with the Bureau of the Budget. However, these trade-offs are given very little consideration in Congress, which continues to be more oriented toward the agency trade-off. Much of the recent interest in Congress in R and D as such is really part of a gradual reorientation toward the new balance of trade-offs being dealt with in the executive branch.

SCIENCE PLANNING AND THE MULTIDIMENSIONAL NATURE OF CHOICE

If we accept the fact that *some* planning has to be done *within* the context of R and D as well as within the context of social purposes, then the question is, what is really meant by planning? It is generally accepted that basic science cannot be planned in detail, except possibly for the construction of large and expensive research tools such as space vehicles, accelerators, or oceanographic research vessels, and the associated supporting organizations. For most of development, and a good deal of applied research, the agency context suffices for planning purposes, although we have seen that this is subject to certain limitations with respect to means-oriented agencies such as AEC or NASA.

The current dialogue between the scientific and political communities is really concerned with just what are appropriate areas for planning *within* R and D. What are the alternatives to be confronted? Planning is inevitably concerned with aggregates, usually expressed as financial obligations for the support of research. These aggregates are the summation of many diverse individual activities characterized by some common property. The politician and the administrator are often bewildered by the categorizations of scientific activity, which are highly overlapping in character. Cosmic rays appear in the guise of space science and also of elementary-particle physics. Astrophysics appears as part of physics and also of astronomy. Materials science overlaps chemistry, physics, and engineering, but as a science often appears as congruent with solid-state physics. Federal contract research centers also employ graduate students and professors and contribute to academic research. What is basic research to the performer may be applied research to the sponsor. Some of the most fundamental research also requires some of the most sophisticated practical engineering. Yet it is precisely the appropriateness of the various possible characterizations of aggregates which lies at the heart of the planning problem and the determination of criteria for choice. Much of the recent literature on the subject has focused on the choice between scientific fields or disciplines, and this orientation of the discussion has tended to be reinforced by the appearance of a series of reports on opportunities and needs of the basic sciences, discipline by discipline, under the auspices of the National Academy's Committee on Science and Public Policy. But the disciplinary categorization is not necessarily the only appropriate one, nor is it reasonable to allocate resources by any single set of categories. The problem of scientific choice is multidimensional.

Whereas development, which constitutes about two thirds of federal support of technical activities, tends to be concentrated in a relatively small number of large projects or programs, research is distributed among literally tens of thousands of individual projects. To collect these projects or expenditures into aggregates only has meaning to the extent that the individual components have some property in common which is significant in relation to some policy goal, whether it be the advancement of a social objective or the development of an institution. For example, it is not at all clear that "basic research" by itself has any significance as a category when it includes activities ranging all the way from a pure mathematician proving a theorem in topology to a huge engineering and logistic effort directed toward a probe to Mars or a Mohole project. Yet these disparate activities are actually included in the category basic research in NSF reports of federal R and D expenditures. Of course, this categorization is legitimate in the sense that all these activities may be considered as motivated primarily by intellectual curiosity, but what policy objective is served by this classification? Is there any meaningful sense in which such disparate activities in basic research should be regarded as competitive with each other, or is the Mars probe more properly competitive with, for example, a national highway beautification program, a subsidy to the merchant marine, or the development of a supersonic transport plane? Almost any simple categorization tends to involve some lumping of apples with oranges.

Among the numerous possible categorizations of research which might have some policy significance are the following:

1. The degree of fundamentality or applicability, for example, basic research, applied research, and development. (In this connection, however, it must be noted that the term "fundamental" is not necessarily opposed to "applied" in research — a misconception which tends to be fostered even by scientists who ought to know better. The term fundamental refers to an intellectual structure, a hierarchy of generality, while the term "applied" refers to a practical objective. It is true that fundamental research is generally less closely related to applications, but not inevitably so.)
2. The scientific discipline, for example, physics, chemistry, biology.
3. The social purpose or function of the research, or its most probable area of practical relevance, for example, health, defense, natural resources, and so on.
4. The institutional character of the research, for example, academic (in universities proper), research institute, industrial, government laboratory, and so on.
5. The scale or style of the research, for example, big science versus little science.

6. The object of study, for example, oceans, atmosphere, space, or materials. Although subjects such as oceanography, meteorology, space science, or materials science are often considered as "disciplines" in the sense of classification 2, they are actually multidisciplinary studies applied to a single class of objects or aspect of the environment.

Of course these classification schemes are not entirely independent of each other either. For example, the more basic or fundamental a research activity is, the more appropriate is a categorization by scientific discipline. Similarly, a disciplinary categorization is much more meaningful for academic research than it is for industrial research or research in government laboratories. The disciplinary categorization is equally limited, and this is becoming increasingly true as advances in understanding have permitted the importation of techniques and concepts from one field of science into another. The growth of science in the postwar era has been characterized by the spectacular growth of hybrid disciplines such as geophysics, geochemistry, biochemistry, chemical physics, computer science, and radio astronomy — or even things as esoteric as neutrino and gamma ray astronomy. Interdisciplinary subjects such as oceanography, atmospheric sciences, and space science draw on all the more classical disciplines, and it is difficult to tell at what point they do or should become disciplines in their own right. Whole areas of research move from one scientific field into another. Atomic spectroscopy, which used to be one of the most active branches of physics, has moved almost entirely into astronomy, although recently it has experienced a revival of interest among physicists under the hybrid title of "laboratory astrophysics." Molecular spectroscopy has become a branch of physical chemistry. New physical techniques such as radiocarbon dating, radioactive tracers, paper and gas chromatography, microwave and nuclear resonance spectroscopy, laser optics, or X-ray diffraction have spread rapidly into many fields of science and technology, at the same time as much of the laboratory instrumentation becomes commercialized. The use of these imported techniques is not passive, however. They have received many contributions and improvements from their adopted sponsors. But the net effect is that it becomes increasingly difficult to define a discipline except by the organizational framework in which it is pursued or the educational background of its exponents; for example, physics is defined as what is currently being done in academic physics departments.

In addition to the problem of overlapping between disciplines, there is the equally difficult issue of how much the support of disciplines as part of other missions should be "counted" in evaluating the total activity in a given discipline. The Physics Survey Committee of the National Academy found, for example, that gross reclassification of NASA expenditures in the early 1960s resulted in radical changes of the total picture for physics and astronomy. Estimates of the support of basic chemistry by the National Institutes of Health, the largest single government supporter of academic chemistry, differed by a factor of two according to the studies made by the Chemistry Survey Committee of the National Academy. This difficulty arises for almost every field where an important mission of a federal agency has been highly dependent on one aspect of a particular scientific discipline, and yet where the overall organization is not discipline-oriented. In many instances the mission-oriented use of a particular discipline may be so specialized that it cannot be regarded as significantly advancing total understanding in the discipline, and this is where the difficulty in categorization arises.

All of these various classifications of research activity warrant some consideration in decisions concerning R and D priorities, or even priorities within research. They are posed in different ways through different means of support and in different federal agencies. For example, choices along the dimension between basic and applied research tend to be posed most clearly within mission-oriented agencies which support the whole gamut of activities from basic research to actual operations. On the other hand, choices in terms of fields of science tend to be posed in the most clear-cut way in the National Science Foundation, where practically all scientific disciplines are covered within a single agency. The choice between different types of research institutions is posed in a more complicated way, but tends to occur indirectly through the choice of agency. Both the National Science Foundation and the National Institutes of Health, for example, tend to be strongly oriented toward academic research, so that budgetary increases for their programs tend to channel into university programs. Other agencies, such as AEC, have strong university programs, but also have developed major responsibilities for the maintenance of broad-spectrum national laboratories. If funds become very tight, the AEC tends to protect these laboratories, while the NSF tends to protect its university programs and hold back on its national centers. Other agencies, such as Interior and Commerce, perform their research mainly intramurally, so that growth of their programs tends to favor certain inhouse laboratories.

Fortunately, modern data-processing techniques lend themselves to the presentation of research support information in many different dimensions simultaneously. We can attribute several different properties to each discrete project according to many different dimensions of categorization. The computer can aggregate these common properties and present them simultaneously in several different ways. This multidimensional mode of presentation is, however, still more potential than actual. We do not yet have the information base on which multidimensional planning can be done.

It is obvious that there is no simple answer to the problem of priorities. It is doubtful, for example, whether any attempt to assign priorities to whole large fields of science such as physics or biology, or even to smaller categories such as solid-state physics or nuclear physics, can be very meaningful. There is too large a variation in the quality and relevance of projects within a broad field to be able to assign a meaningful priority to a field as a whole. This is somewhat easier with respect to big science than little science because the decisions to be made involve sufficiently large amounts of money and are sufficiently few in number so that they can be sharply focused in the budgetary process, but even here the other dimensions of categorization must be taken into account. The disciplinary reports of the Committee on Science and Public Policy of the National Academy have made an important contribution to the discussion of the priority problem, but they do not and cannot address themselves directly to the problem of priorities, except within fairly homogeneous subfields. Each discipline has attempted to put its best foot forward and to utilize those arguments which present its case most strongly. Thus chemistry and plant sciences have relied heavily on the usefulness of their basic results to industry and agriculture, while astronomy has appealed to the great fundamental and philosophical importance of its investigations. How is one to make an effective comparison between these two justifications? The difficulty is compounded by the fact that immediate usefulness may not be the same as greatest usefulness in the long run. Although it may take much longer for the results of basic physics investigations to filter into technology, because they often do so *via* other sciences rather than directly, the net impact of basic physics could be greater in the long run. Past history suggests that this is so, but even here history is not necessarily a reliable guide. Intellectual structures and styles are changing too rapidly. One of the unfortunate side effects of demanding too immediate a payoff from a basic research field is that it tends to lead the practitioners into a species of intellectual dishonesty not only with respect to the public and the politicians but even in their own minds. Sound scientific judgment can be corrupted by too much preoccupation with early payoff. This is partly what Weinberg means when he refers to the need of a general overhead type of basic research to supply "style and sophistication" to the whole technical enterprise. Strong basic research is the guarantor of intellectual integrity in applied research and development.

Of the $15 billion federal budget for "science" at least two thirds is for development, that is, it is sharply focused on the time-bound achievement of certain specific end products or concrete certifiable objectives, such as landing a man on the moon. Clearly this part of "science" not only can but must be planned. In fact, one of the major achievements of the last ten years has consisted in advances in the state of the art for planned achievement of large, well-defined technical objectives in a rational and stepwise fashion. This art is most fully developed in areas where the limiting factors in the achievement of objectives are primarily technical, such as defense and space. In areas where social and economic factors are limiting, or at least a major component, and where large numbers of lay people have to be convinced of the validity of the specific steps of a plan, the planning art is much less well advanced, and the criteria of success much less clear-cut. We do not yet have the right tools for planning the attack on major "civil" problems such as environmental pollution. In fact, we do not know how to translate the global problem into a series of manageable technical problems, or realizable finite technical objectives, which is really the essence of good planning.

When it comes to science proper, in contrast to the highly structured utilization of science, the nature of planning changes entirely. Here it becomes a matter of providing over a sustained period the best possible environment for decentralized initiative, and for individual and institutional creativity. This does not necessarily imply the complete absence of choice, or the delegation of the problem of choice entirely to the scientist in the laboratory. Part of the "environment" must consist in the posing and formulation of goals which are both scientifically challenging and socially rewarding. The task of formulating such goals in the broadest terms and making them appear attractive, interesting, and realizable to individual scientists and technologists is really the major task of scientific "planning." But we must also be prepared to accept the fact that some goals, no matter how important or rewarding they may appear to the layman, are just not ripe for scientific attack, while other scientific goals, which appear remote and esoteric to the layman, may be ripe for attack and may in the long run lead to practical results more economically and sooner than more apparently "practical" research. The best measure of whether a field is ripe for attack may be its ability to attract first-rate scientists. We must be prepared not to force the system too hard by trying to use money to induce scientists to "push worms into the ground," as one of my colleagues graphically describes it. What I am trying to say is that publicizing and debating scientific goals among scientists and laymen is itself a mode of scientific planning, in that it serves to draw the younger uncommitted new scientists into important new fields. Planning in this sense is a basically democratic process, a cultivation of a climate of opinion among scientists rather than a coercion of the system. The alternatives of a wholly autonomous science on the one hand, or a scientific enterprise regulated wholly by directives from the top on the other, are unrealistic. What is best both for science and for society is a system which provides a large measure of resources for scientists to do what they think most important and significant in terms of their own criteria, but through constant publicity, analysis, and involvement of the scientific community in the decision process, directs the attention of scientists to the problems which society thinks are most important and at the same time helps in the reformulation of society's problems in scientific terms. At the same time, a science which is wholly directed and governed by practical objectives is likely to lose its soul and to cease to attract the best minds. While it is true that basic science tends to enjoy the highest intellectual prestige, it is also true that economic forces and the natural overproduction of Ph.D.'s in basic science are continually seeding applied science and technology with new people and hence new viewpoints, techniques, and ways of asking questions. The most important factor in the translation of basic science into practical results is the continuing interchange of people between science and technology.

With respect to the continuing tension between "mission-oriented" and "science-oriented" research, I believe that in itself this tension is a creative force in our society, and that its preservation should be made a major objective of policy. In particular, I believe the time has come when some of the great research institutions we have built up primarily as the captive of particular missions and particular agencies should be regarded more as a national resource and should be used for the solution of national problems with less consideration than at present of the mission of the particular agency within which they happen to lie.

THREE □ Can Science Be Planned?

The following chapter is a reprint of a paper prepared in connection with an international conference on science policy held under the auspices of the OECD at Jouy-en-Josas, near Paris, on February 21 to 25, 1966. The paper is an attempt to set forth in a systematic manner the various possible rationales for the support of basic science by governments. As compared with the preceding chapter on the allocation of resources, it deals more with the general philosophy and less with the practical politics of resource allocation for science.

INTRODUCTION

In a sense the title question is a rhetorical one. Science *is* planned, whether implicitly and largely as a by-product of decision-making processes external to science, or explicitly and consciously with respect to science and technology themselves. People who write about planning of or for science usually seem to be talking primarily about fundamental science, and the planning process is thought of in terms of relative governmental funding of different scientific disciplines. In fact this is a grossly oversimplified view of planning, because in fact the scientific discipline is only one of many dimensions of planning, and probably the one least susceptible to nonexpert judgment.

The real issues involved in scientific planning mostly relate to how to reach the best adjustment between the need of science for internal autonomy and the desires of society for the fruits of science. At one level these are mutually conflicting requirements, but at a higher level and on a longer-time scale they are probably mutually supporting, although the extent of their conflict and mutual support is, in fact, what much of the debate is about.

The problem has been succinctly summarized by Barber[1] as follows: "However much pure science may eventually be applied to some other social purpose than the construction of conceptual

[1] B. Barber, *Science and the Social Order*, Collier Books, New York (1962), p. 139.

schemes for their own sake, its autonomy in whatever run of time is required for this latter purpose, is the essential condition of any long run applied effects it may have." Yet it is also true that the social effects of science do not follow automatically from its existence. There are many instances, both contemporary and historical, where eminence in pure science and success in technological application are not found in the same nation or the same organization. The translation of scientific discoveries into socially useful applications is generally a more difficult and lengthy process than the transfer of science or technology, once developed, between institutions or nations of comparable scientific sophistication. Thus a nation or organization which has the managerial and technical capability to translate new discoveries or inventions quickly into successful commercial products or public technology may reap the benefits sooner and more effectively than the originator.

Thus the sorts of questions at issue in the planning of science are the following:

To what extent should planning be left to the internal workings of the scientific system, and to what extent and in what degree of detail should it be imposed from outside science itself, using external criteria related to applications?

Who should be involved in the setting of scientific goals at various levels of detail? While the fine structure of scientific planning must be done largely by scientists themselves, societal judgments are involved at increasing levels of generality; how are societal judgments to be melded into the process of decision-making?

How is the system of scientific choice to be institutionalized? For example, to what extent should peer judgments be used in selecting projects for support, and how is the objectivity of the selection process to be assured?

What measures can be made of the validity of past planning or the effectiveness of project selection?

What are the categories of classification of scientific activities for which it is appropriate to plan? For example, should relative allocations between basic research, applied research, and development be determined independently for each broad mission, or is there some optimal global allocation to these categories that should provide a guide to overall planning of research and development, for example across the whole federal budget? Is there an *a priori* preferable allocation of resources between universities, industry, and government, or is this to be determined on a mission-by-mission basis?

For many purposes planning can be related to scheduling, that is, to the time horizon of research, and for this purpose it is clear that the research needed in the development of a specific scheduled system must be much more closely planned than research which is not aimed at filling specific gaps in knowledge to fit a time schedule. However, this does not really settle matters, for the question still arises whether it is better to tie as much research as possible to a specific desired end result, or whether one should allocate to a given end result only the minimum of research necessary, and support the rest on a freer and more generalized basis? To put the matter another way, to what extent should research be planned to advance the general state of the art, and to what extent should it be planned to achieve specific systems?

CATEGORIES

My own feeling is that the categorization of research for planning purposes is most easily done in an institutional context. That is, research is most easily classified and planned in terms of the major purposes of the institutions which conduct it. Such a classification is much more meaningful than division into scientific disciplines or into generalized social missions. For example, it is not very useful in policy terms to classify both a

Mariner probe to Mars and a 36-inch telescope for a small Midwestern college in a global allocation for astronomy. Nor is it much more meaningful to lump together the research and development for a military logistics aircraft with research and development aimed at a non-air-polluting bus for intracity transportation as transportation research.

For purposes of planning I have suggested three broad institutional categories as follows:[2]

1. Mission-oriented or nonacademic research, where the term "mission" refers to an objective defined in societal rather than technical terms. The mission-oriented research establishments are vertically integrated organizations which usually span a broad range from basic research to development and even technical support of manufacturing or operations. Such organizations may be run directly by government or be operated under contract to government by universities, private industries, or nonprofit corporate bodies, or they may be funded entirely or partially from private sources, either within industry as industrial laboratories, or as independent research institutes. Although they may conduct the most sophisticated and fundamental types of research, their purpose is primarily nonscientific; their scientific activities are a means to some other purpose of the institution. In such organizations the division among disciplines and the division between basic and applied activities are matters primarily internal to the organization itself, within its overall funding, which depends on the social importance of its mission, its long-range success in achieving its goals, and the various resources on which it can draw.

[2] H. Brooks, "The Future Growth of Academic Research: Criteria and Needs," Hearings before the Subcommittee on Science, Research, and Development, 89th Congress, June 1965 (Washington, D.C.: U.S. Government Printing Office, 1965), I, pp. 362–363.

2. Non-mission-oriented, but institutional basic research. The organization we have in mind is typically the research institute whose mission is defined primarily in scientific terms, for example high-energy physics, or molecular biology, or geophysics. Although its individual scientists may have a great deal of freedom, such an institution follows some sort of coherent program constantly adapted to the changing frontiers in its area of interest. It is usually characterized by a much larger ratio of supporting staff to independent scientists than in the typical academic research setting discussed below. It is often, but not always, characterized by expensive and complex facilities requiring long advance planning. In the United States such an organization has typically been created to serve the needs of academic science, and a substantial portion of its activities is under the direction of visiting scientists from universities or other institutions.
3. Academic research. This is usually small-scale research carried out in academic departments by students or other short-term apprentices under the direction of senior staff who also teach. The planning required is usually short-range and the supporting staff minimal in comparison with either categories 1 or 2 above. What I have called academic research usually corresponds roughly but not completely to what is referred to as "little science," while institutional research more often involves "big science." On the other hand, both institutional and mission-oriented research may involve substantial activities which are, considered by themselves, indistinguishable from academic research.

The basic idea behind the above categorization is that the character of research and the way its results are used depend most strongly on the institutional environment in which it is conducted. This does not mean that the three types of research listed are not symbiotic; they are. In fact, the success of their symbiosis depends partly on the degree to which their activities are overlapping, since it is through this overlap that communication between the various types of research activity occurs. A healthy development of all three categories is necessary to a healthy national or international scientific effort. It is also true that, like almost all categorizations of science, this one is an oversimplification. One can certainly identify a complete spectrum of organizations with characteristics shared between the categories. For example, there are centers of high-energy physics on university campuses which have all the characteristics of institutional science, except that they are more involved with professors and graduate students. There are large national laboratories, notably the Argonne Laboratory in the United States, which combine activities in categories 1 and 2 above. The Max Planck Institutes in Germany are to some extent mixtures of categories 2 and 3. In the United States it is regarded as highly desirable that institutions in category 2 be as closely associated with universities and graduate education as other circumstances permit. In fact, there is a fairly widespread belief in the United States that the absence of sharp divisions between categories, and the resulting mobility of scientists and engineers between different types of institutions and categories of research activity, are important facilitating factors in the rapid conversion of scientific findings to practical application. The short-range benefits of this mobility were striking during World War II, but have perhaps been somewhat lost as institutional patterns have tended gradually to solidify.

THEORIES OF SCIENCE PLANNING

In the literature of science planning one may distinguish several theories which may be roughly classified into four types as follows: (1) science as an autonomous enterprise, (2) science as a technical overhead on social goals, (3) science as a social overhead investment, and (4) science as a consumer good. I will discuss each of these in turn.

The first view is often identified with the name of Michael Polanyi, who viewed the system of science as an entirely self-regulating community, much like the economic model of the laissez-faire school. According to this view any social interference with the autonomous workings of science would slow scientific progress and thus also delay the realization of the benefits of science which the social intervention was designed to achieve. The Polanyi school claimed freedom of science, not as an inalienable right of scientists, but as a means to maximizing the efficiency of the scientific system. In the same way the laissez-faire economists claimed the freedom of enterprise and private property not as an inherent right, but as the best road to economic efficiency. In each case there was the implicit assumption that maximum scientific growth or maximum economic growth, as the case might be, would automatically maximize all other desirable social goals.

The Polanyi thesis was to some extent a reaction against the Marxist view that science was entirely determined by economic needs and current technology. The Polanyi view has had a much stronger influence on academic science in Europe than in the United States. In fact, from its beginning the United States has taken a highly pragmatic view of science both in education and government, though not on theoretical grounds of economic determinism.

An interesting variant of the Polanyi thesis has been put forward by Derek Price.[3] Whereas the Polanyi school tended to regard science as a delicate plant, which would wilt if interfered with by society, Price seems to regard science as a vigorous weed that society could not interfere with even if it desired. Price would not deny that nations have occasionally committed scientific suicide, like Germany in the 1930s, but he would argue

[3] D. J. de S. Price, "The Science of Scientists," *Medical Opinion Review*, 10 (1966), pp. 88–97.

that this is a transient and unimportant perturbation of *world* science, which grows and advances according to its own inexorable laws of internal development. The development of this worldwide system is not much affected by local policies because scientific knowledge has such a high rate of diffusion.

Price draws a sharp distinction between science and technology, but regards them both as cumulative systems, each developing relatively independently of the other, though with some sort of complex and little understood symbiosis.[4] He also points out that, as judged by the output of literature in more or less fundamental science, each nation contributes to the literature in proportion to its GNP, but there is much wider variation in investment in technology relative to GNP, the United States and the U.S.S.R. each spending about twice as much, relatively, on technology as Western Europe.

According to the Price view there is no true problem of "planning" in science, as opposed to development. By "science" in this context Price means any scientific research whose results are published in the scientific literature and included in one of the world's major abstract journals. Thus it includes a good deal of research ordinarily classified as applied. Price shows that the relative distribution of abstracts among fields of science varies but little among the industrial nations, and from this he concludes that relative allocations among fields are primarily a function of the world communication pattern of science, little subject to social control in the long run. He goes on to suggest, in fact, that national planning for science should consist largely in matching the national effort to the world pattern, that the primary effect of reallocation of resources may be to affect only the quality and significance of the work but not its volume, or to induce migrations of scientists.

In understanding the autonomy theory of science it is im-

[4] D. J. de S. Price, "Is Technology Historically Independent of Science? A Study in Statistical Historiography," *Technology and Culture*, 6 (1965), pp. 553–568.

portant to distinguish between freedom and autonomy. In this school of thought science claims autonomy in the management of its own growth and development, but the activity is actually a highly disciplined one with a well-formalized system of sanctions and rewards. The rewards give maximum recognition to originality and conceptual generality in a man's contributions to science, while there are strong sanctions against spurious claims of originality or failure to acknowledge and document all prior work which is relevant. Thus, although the individual scientist regards himself as completely free to choose his own line of work, his choices at any one time are heavily conditioned by his training, his previous research, and the contemporary contributions of colleagues all over the world. That this constraint is real is suggested by the great frequency with which similar or even identical discoveries or concepts appear simultaneously all over the world. This could scarcely occur in a system as random or arbitrary in the relationship of its individual components as the use of the word scientific freedom might suggest.

The second principal view, expressed most articulately by Weinberg,[5] is that of science as a technical overhead on other social goals. In this scheme society is viewed as allocating resources primarily to technology in the broad sense of developing or improving capabilities for the achievement of certain social goals. The allocations are made to agencies or institutions responsible for certain social missions, and these in turn devote a certain proportion of their allocation to the support of science which is thought to contribute in some measure to the mission. Weinberg leaves unanswered the question of what proportion of agency funds should be devoted directly to technology and what to science, but implies that this is a decision which should be made at many different levels. Each level of management is

[5] A. M. Weinberg, "Science, Choice, and Human Values," *Bulletin of the Atomic Scientists,* 22 (1966), pp. 8–13.

thus expected to set aside a certain proportion of its allocation for general-purpose research, that is, for research having objectives beyond an immediate time-bound developmental goal. Thus an agency head may allocate his resources among laboratories, each one of which may then devote a portion of its funds to research. However, the agency head also reserves a proportion of his total allocation to support research, for example, in universities, and finally the chief of state, or his budgetary officers, set aside a small fraction of the total national budget for research of the most fundamental and least applicable character, primarily in order to establish standards of quality and set the style for the remainder of the national scientific enterprise. This allocation model applies equally well, of course, to the private sector. In a large corporation, for example, the missions correspond to the operating divisions, each of which has its own research activity, with considerable local autonomy within its allocation, but there is also a "corporate laboratory" which performs the most fundamental and generalized research for the whole company.

Thus, in the Weinberg model, the process of allocation is a rather decentralized one, with science appearing at many different levels of management as one of several means or tools available for the execution of its particular subtask in the overall mission of the organization. In Weinberg's scheme there is no "science budget" in a global sense, only a series of science budgets attached to particular social missions, with the "corporate research budget" — the only true science budget — being a relatively small proportion of the total expenditure for science by the organization. The Weinberg model has the great virtue of being a fairly accurate and realistic description of how science budgeting is actually structured in the United States. In this scheme the global overhead would appear as the research funds available to the National Science Foundation, which constitute only about 15 percent of the total government funds for academic research and less than 5 percent of all government research expenditures (excluding development). There is, of course, an anomaly here, in that the NSF is merely one agency among equals, the others being "mission-oriented," whereas in Weinberg's model it should be attached to the President. On the other hand, it is, perhaps, no accident that the White House has had to pay special attention in the past to the establishment of the level of the NSF research budget. The situation is further confused by the fact that the NSF does have a "mission" in terms of scientific education, and certainly some of its research money must be regarded as attached to this educational mission, particularly in relation to graduate education.

However, the fundamental difficulty with the Weinberg model is that it exaggerates the clarity of the distinction between means and ends. At least this criticism applies if the model is to be viewed as a description of reality, rather than a prescription for policy. In the first place, two of the major government spenders for research, the AEC and NASA, have missions which are to exploit areas of technology rather than to achieve goals which would normally be described as societal. To some extent, it is true, the very existence of such agencies on the same level with the Cabinet departments implies that these particular technologies have been elevated to the status of social goals in their own right. They have become ends rather than means essentially by political definition. There are obvious historical reasons for special treatment of these two technologies, and they have been similarly treated in essence in all the advanced nations. Nevertheless, from the standpoint of resource allocation, it is difficult to compare atomic energy with pollution control, or space with the removal of urban blight, as is evident from the sorts of debate now going on in the United States. The existence of agencies defined by technology creates a vested interest in particular technologies — in particular means — which tends to make them self-perpetuating in relation to other social objectives. Such technology agencies do not really conform to Weinberg's description.

On the other hand, there is also a problem if Weinberg's model is taken not as a description but as a prescription for policy. The fundamental issue here is really the issue which was debated so vigorously when the first Atomic Energy Act was written in the United States. Both atomic energy and space could have been developed entirely under military auspices, but the question arises whether their other social benefits could have been achieved if they had been developed in this exclusively "mission-oriented" way. The proponents of separate agencies argued that leaving development to the military would automatically give low priority to other potentially more important civilian applications of nuclear physics and technology or of space science and technology. The inadequate support given to the satellite project in the United States, while rocket technology was exclusively under military sponsorship, is cited as one example of the inadequacy of mission-oriented support, at least in the early stages of a new technology. At such a stage the full potentialities of the new knowledge are never sufficiently visible to justify adequate support by a mission-oriented agency with other competing responsibilities. Similarly it is argued that the extensive applications of radioactive isotopes in medicine and agriculture would never have been developed by the military, on the one hand, or the Department of Agriculture or Health, Education, and Welfare on the other, until the groundwork of technology and research had been laid by an agency focused on the technology itself rather than the component social missions. Perhaps, the AEC and NASA can be fitted into Weinberg's conceptual scheme after the fact, but it is doubtful whether the elevation of atomic energy and space to social missions in their own right could ever have been fitted prescriptively into this conception.

There is now a national debate going on in the United States with respect to the handling of two of the environmental sciences, oceanography and the atmospheric sciences. Both have hitherto been handled under the mission-oriented prescription, with responsibility split among many agencies and coordinated through special committees in the Office of Science and Technology. But many are now arguing that these subjects ought to be gathered either into a single environmental sciences and technology agency[6] or into two agencies. In the report, "Effective Use of the Sea," issued under the auspices of the President's Science Advisory Committee, the argument for a single overall agency is explicitly framed in terms of the commonality of the *means* used in all the environmental sciences, and this coherence of concepts and techniques is asserted to outweigh the diversity of separate missions. Only after the workability of certain operations such as weather modification or marine farming has been demonstrated by the specialized agency is it regarded as safe to turn over the further development to a mission-oriented agency.

Furthermore, the situation presented by atomic energy and space, and by the environmental sciences, is likely to become increasingly common. As stated by Daniel Bell,[7] "what has now become decisive for society is the new centrality of *theoretical* knowledge, the primacy of theory over empiricism, and the codification of knowledge into abstract systems of symbols that can be translated into many different and varied circumstances." Thus the commonality of conceptual structure is becoming increasingly important as a determinant of organization in comparison with commonality of mission. In Weinberg's own terms,[8] the whole problem of innovation and growth may be

[6] Panel on Oceanography, President's Science Advisory Committee, "Effective Use of the Sea," The White House, June 1966 (Washington, D.C.: U.S. Government Printing Office, 1966), pp. 88–90.
[7] D. Bell, "Notes on the Post-Industrial Society I," *The Public Interest*, 6 (Winter 1967), pp. 24–35.
[8] A. M. Weinberg, "But Is the Teacher also a Citizen?," *Science*, 149 (1965), pp. 601–606.

becoming increasingly "discipline-oriented" rather than "mission-oriented."

In actual fact it may be possible to arrive at an intermediate position. It seems clear that a new technology tends to be "discipline-oriented" in its early phases, while it becomes increasingly "mission-oriented" as it matures and the scope and limitations of its possible social applications become clear. What is, perhaps, needed is a management system which provides a finite lifetime for technological agencies, with provision for transfer or "purchase" of the technology by user agencies at a certain stage of development, together with considerable transfer of personnel. To some extent one can already trace such a pattern in the United States and, to a lesser extent, in other countries. In the space field, for example, the Commerce Department is assuming responsibility for specification, acquisition, and management of a weather satellite system. A mixed public/private corporation, COMSAT, was created to take over similar functions with respect to satellite communications. It has been proposed that a new international agency be created to take over satellite surveillance and to disseminate the intelligence thus acquired as a measure of international pacification — reducing the fear of military surprise. In the atomic energy field, the production and development of the "mature" reactor technologies such as boiling and pressurized water have been taken over by the private sector in the United States with active encouragement from the government, and the Central Electricity Authority has taken over a major part of civilian reactor technology in the United Kingdom. The development of nuclear propulsion for ships has been taken over by the Navy and the Maritime Administration, respectively, for military and merchant ship propulsion. With the introduction of program budgeting in the Defense Department the cost of nuclear weapons is now assessed against the corresponding complete weapons systems. In many of these examples the technology agency, originally supported for its own sake, finds itself increasingly in the position of vendor to the purveyors of other social missions, especially in cases where common facilities such as launch vehicles and tracking networks may be needed for several different missions.

The problem still remains of the tendency of technologically oriented agencies and institutions to outlive their original purpose, especially in the absence of a market mechanism to determine their optimum size. One suggestion which has been made is to create a single general-purpose technology agency which would serve as a common home for new technology in its early phases, but would have a clear mission to transfer the technology, along with part of the associated personnel, to other mission-oriented organizations, including the private sector, as quickly as possible. One possibility would be to transform existing atomic energy agencies into such agencies, since they already have great technological breadth, and tend to be running out of their original innovative mission. One sees small trends in this direction in both the United States and the United Kingdom. A somewhat similar role is apparently envisioned for NSF in legislation recently introduced in Congress, and in fact the NSF has played such a role in the past in the field of weather modification, which is in the process of being partly transferred to operating agencies. Such a role for NSF in applied science would also be consistent with its educational mission — the initiation and transfer of new technology being conceived as a kind of educational role.

The third view of science and its role in society follows rather naturally from the viewpoint expressed in the preceding paragraph. It is the view of science as a social overhead investment. In this view science is still an overhead in the sense that specific activities cannot be uniquely associated with particular social purposes, but a much larger proportion of it is regarded as underlying *all* the purposes of society, and is therefore to be carried out in an organizational structure which is patterned on the conceptual structure of knowledge. In economic parlance a social overhead investment is an investment in the sense that it makes possible the productivity of a society but an overhead in the sense that it does not contribute uniquely to any particular aspect of productivity. Thus, in this view, the support of science has very much the same social function as the support of education, and in fact at the more advanced level science research and education grade continuously into each other. Even the most advanced and sophisticated forms of basic research may be regarded partly as maintaining and developing a pool of highly skilled manpower and of fundamental knowledge which may on occasion be deployed for more applied social purposes. "Every society," says Daniel Bell,[9] "now lives by innovation and growth, and it is theoretical knowledge that has become the matrix of innovation." But the matrix to which Bell refers cannot be considered only as codified information. It is a living organism which, in a sense, can only be maintained by continuous self-renewal carried on by a community of people, who are constantly sharpening their perceptions and insights. No modern society can afford to do less than develop its potential talent to the maximum degree possible. The development of talent, moreover, involves not only education in the usual sense of the diffusion of knowledge and understanding, but also the constant acquisition of new insight and understanding by those who have the talent to advance the frontiers. When men reach the point where they can no longer learn from books and teachers, they turn to nature itself as their teacher, but research is still no less a learning process than what is involved in more formal education. The social overhead view of science considers the human resource aspect as of equal importance with the scientific results themselves. And talent goes on pyramiding throughout the basic research process.

[9] Bell, *loc. cit.*

As pointed out by Ben David, the consideration of education and research as a social investment, that is, as an investment for future social productivity, is relatively recent. In almost all countries it has been regarded as primarily a benefit to the individual rather than to society, a privilege open primarily to an elite. Even in countries like the United States, where mass public education was introduced early, the arguments for it were made mainly in terms of social justice in a democracy, that is, equal opportunity to the individual as part of his "inalienable right" in a democratic society.

The fourth view of the social role of science may be termed science as a consumer good. In this view science becomes just one of the many ways society expends its excess product, and is not different in kind from any other form of cultural expression such as art, music, or imaginative literature. Of course, humanistic culture can also be viewed to some extent as an overhead in the Weinberg sense, setting the general tone and standards of the whole society, but its connection with the productivity of the society is probably harder to establish than in the case of basic science.

One may, perhaps, distinguish two variants of the consumption good view of science. The first was articulated by H. Johnson in his paper in the National Academy study on "Basic Research and National Goals." [10] In this view basic science is purely and simply a luxury good, accessible only to the very few privileged people in a society who have the education to appreciate its esoteric mysteries, and who have been able to persuade a sufficient number of their fellow citizens to support their activities. The more affluent a society, the more nonproductive activities of this sort it can afford. The American space program is only an extreme example, somewhat analogous to

[10] H. G. Johnson, "Federal Support of Basic Research: Some Economic Issues," in "Basic Research and National Goals," A Report to the Committee on Science and Astronauts (Washington, D.C.: U.S. Government Printing Office, 1965), pp. 127–142.

the conspicuous consumption of the *nouveau riche* of the early twentieth century. In the Johnson view there is no more reason, nor less reason, for spending public funds on science than on artistic and humanistic activities. Science may add a certain amount of elegance and style to socially necessary technology, just as architecture adds style and elegance to public buildings, but it is neither more nor less functional than architectural design or elegant interior appointments.

A modified and somewhat more sophisticated view has been put forward by Stephen Toulmin,[11] who suggests that science in a modern society is a part of the basic purpose of the society. With the advent of the affluent industrial society science tends to replace economic productivity as a primary social goal. As society is able more and more to satisfy its material needs with less human effort, it becomes more preoccupied with its spiritual and intellectual needs. It must develop new goals and aspirations in order to remain viable as a social organism. Of all the nonmaterial activities of society, science is the most open-ended and cumulative. One of its characteristics is that each new question which is successfully answered opens up several new questions in its place and often reveals new worlds for investigation. Whether this will always be so is a matter for speculation, especially in this era of apparently accelerating progress. Many times in the past scientists have believed that all the significant questions had been answered, and the only task remaining was to fill in the details, to work out the full ramifications of a conceptual structure whose main framework was completely delineated. Yet each time this expectation has proved to be wrong. Each new major advance has revealed an unsuspected new world, a new conceptual structure embedded in the old, and subsuming it. The quantum physics of atoms and molecules subsumed the mechanics of Newton, and was in turn subsumed

[11] S. Toulmin, "The Complexity of Scientific Choice, II: Culture, Overheads or Tertiary Industry?," *Minerva*, 4 (Winter 1966), pp. 155–169.

by the nuclear physics of Rutherford and Bohr, which in turn is subsumed in the elementary-particle physics of the modern era. In a similar manner the evolutionary biology of the nineteenth century is subsumed, apparently, in the molecular biology of the twentieth. It is also true that, while the discovery of these more "fundamental" worlds embedded within the old may have temporarily diverted the interest of science, the macrocosms themselves experience periodic revivals, as new experimental and conceptual tools for dealing with complexity are discovered.

CRITIQUE OF THE VARIOUS VIEWS

To some extent these views are not necessarily mutually exclusive and may even be regarded as complementary. They emphasize different aspects of science, and different activities are justified more in terms of one view than another. It is probably rather difficult, for example, to justify the study of quasars or of neutrino astronomy in terms of a social overhead investment or even technical overhead. It is true that some of the most fundamental investigations in astronomy and elementary-particle physics also place demands on advanced technology which in turn lead to developments which can be adapted to other uses of social importance. For example, high-vacuum techniques and techniques for radioactive counting have been developed in connection with nuclear research and have later found many other applications, so that even the most esoteric research can be at least partially justified retrospectively in terms of applications more directly related to society. However, it is rather difficult to justify the timing and relative magnitude of such research investments *primarily* in terms of such by-product benefits.

Unfortunately, none of the four views articulated above gives much basis for quantitative criteria of planning. The Derek Price view and the social overhead investment theory probably come closest to providing a quantitative basis in that both imply that the total support of science should be such as to engage all the available talent. There are two difficulties even with this approach. In the first place it applies mainly to what I have termed academic science. As I have pointed out elsewhere,[12] nonacademic science, or institutional basic research, tends to be open-ended in economic terms. A single space scientist or oceanographer can command vast resources of supporting talent and industrial products, and even within academic science the availability of increased financial support makes it possible to substitute the purchase of commercial instrumentation for the construction of homemade apparatus. Thus even "little science" tends to become open-ended in its capacity to absorb resources if increasing support is sustained over a long enough period, so that industrial development can catch up to the induced demand. In the second place there is considerable uncertainty as to how large the pool of talent for science really is. On the one hand, it is estimated that we are probably exploiting less than 10 percent of the talent potentially available for science and technology. For various reasons of motivation, social background, availability of educational opportunity, etc., many people do not go into science. The United States channels a considerably higher proportion of each age cohort into science and engineering than any other country in the world,[13] but the poor quality of elementary and secondary science education suggests that many more potential people are lost to science. Other nonscientific activities require high talent also, especially managerial talent, and it is sometimes argued that support of research has

[12] H. Brooks, "Future Needs for the Support of Basic Research," in "Basic Research and National Goals," A Report to the Committee on Science and Astronautics (Washington, D.C.: U.S. Government Printing Office, 1965), pp. 77–110.

[13] W. R. Brode, "Who Speaks for Science," in "Topic 11: New Horizons for the Physical Sciences," *A Journal of the Liberal Arts*, 6 (1966), pp. 14–34; W. R. Brode, "Approaching Ceilings in the Supply of Scientific Manpower," *Science*, 143 (1964), pp. 313–324.

diverted or will divert needed talent from these other activities. On the other hand, analysis of the actual trends in types of degrees granted both at the baccalaureate and the doctorate level make it quite clear that there has been almost no shift of talent between the broad areas of natural science and other areas in the last sixty years.[14] Furthermore, studies of the publication habits of scientists after the doctorate indicate very little change in the percentage of doctorates who publish anything other than their Ph.D. thesis (about 20 percent). Except for compensated fluctuations due to the two wars, the scientific "community" has grown at a constant rate of about 6 percent annually, and this growth is only marginally faster than other professions requiring comparable educational background. One might thus be inclined to argue that support for science ought to increase at a rate sufficiently rapid to fully utilize a fixed proportion of available intellectual talent, but no faster. This would be entirely consistent with the social overhead investment model.

Even if one were to accept the above arguments for the scientific enterprise as a whole, there are problems of allocation of talent within science, not only between fields of science, but also between fundamental and applied activities. The Price-Polanyi view suggests that there are strong forces embedded in the internal ecology of science which tend to determine these relative allocations, and that the primary task of the allocation mechanism, especially as it pertains to the less mission-oriented parts of science, is to discover what this natural distribution among disciplines and subdisciplines is. Most administrative mechanisms for the allocation of resources to university science reflect this philosophy in one way or another. For example, in the United States the philosophy is reflected in the widespread use of the project grant system, with proposals of research from small scientific groups led by "principal investigators" selected

[14] H. Brooks, "The Future Growth of Academic Research: Criteria and Needs," *op. cit.*

by "juries of scientific peers." No matter what the announced philosophy of the supporting agency, peer judgments tend to reflect primarily internal scientific criteria. In mission-oriented agencies which use peer judgments, such as NIH or the Army Research Office, criteria of mission relevance are also applied in selection, but in practice the application of this criterion tends to take place through the selection by the investigator of what agency or subdivision to apply to rather than through the reviewing panel itself. The tendency of the panels is to interpret mission relevance as liberally as possible, and to use it sharply only when comparing proposals of approximately equal scientific merit. Such a group will seldom reject a remotely relevant proposal of very high scientific merit in favor of an average or mediocre proposal of greater relevance. In Europe and the United Kingdom, where academic research funds tend to be allocated more through institutional grants, scientific merit tends to become an important criterion of internal allocation within the institution. University appointments the world over are determined primarily in terms of excellence in a particular scientific discipline as judged by reputation among peers in the same field.

The United States's system of project grants, however, probably provides considerably greater differential support for scientific excellence and high scientific productivity than the institutional support system more commonly used in Europe. Where funds are allocated internally within the university, the essentially democratic and political character of university administration generates strong pressures for equal distribution of support, regardless of merit, especially to senior people whose earlier productivity may have declined. Even with the project system, the United States is beginning to realize that similar pressures operate for the nation as a whole, as indicated by demand for more equal geographical distribution. The same phenomenon is reflected in Europe by an increasing tendency to question the financing of international research centers. These "equalization" tendencies are probably bad from the strict standpoint of scientific efficiency, that is, the best science for the least money, but there may be certain compensating advantages in the wider distribution of advanced education and a quicker rate of diffusion of new science and technology. The advantages of this diffusion in certain cases are well illustrated by the case of agricultural research in the United States, where the land-grant college and agricultural experiment station system have made United States agriculture one of the great technological success stories of all time.

In practice the project grant system in the United States seems to have achieved a moderately broad institutional distribution but with considerable concentration on the most productive units *within* institutions. This is particularly well shown by the statistics given in the Westheimer report on chemistry.[15] A somewhat similar effect seems to be achieved by the research councils in the United Kingdom, by CNRS in France, and by the Max Planck Institutes in Germany.

The purest application of the Price-Polanyi philosophy is the attempt to allocate resources among disciplines on the basis of "proposal pressure." Although this criterion is only approximated in practice, the frequency with which proposal pressure is used as an argument for more funds in a particular field reflects an underlying assumption of the philosophy.

Ideally it plays the same role with respect to whole disciplines that evaluation of individual project grants does with respect to individual research groups. Proposal pressure involves, in principle, support of the same proportion of scientifically approved proposals in each discipline, although in practice allowance is made for the fact that some proposals are discouraged informally

[15] National Academy of Sciences-National Research Council, *Chemistry: Opportunities and Needs* (Washington, D.C.: U.S. Government Printing Office, 1965), p. 206.

and some program officers attempt to inflate proposal pressure by soliciting proposals even in the absence of funding prospects. It is no accident that proposal pressure is most frequently used in NSF which is the least mission-oriented of the agencies, at least with respect to the *content* of the science it supports. A problem with proposal pressure is that, since evaluation is done by disciplinary specialists, there is not a very strong mechanism for comparing the quality cutoff of various fields, although to some extent this is done subjectively by government program officers who may be broadly knowledgeable about several fields.

If one examines the actual facts of resource allocation for science and technology a rather interesting model suggests itself, which seems to conform closely with the Price picture although it has some overtures of Weinberg. Each nation allocates all it can "afford" to *technology*. Here the term "afford" applies to both physical inputs and trained manpower. Beyond this it supports academic type research — the kind whose output is papers published in recognized journals and appearing in the abstract literature — as a *fixed* overhead, not on applied work, as in the Weinberg model, but on its total GNP. It is interesting that this overhead is related to output rather than input of science; it seems to be automatically adjusted for the difference in cost of science relative to GNP. Nations with a higher applied investment apparently have a higher cost of academic type research largely because of competition for inputs. Academic research has to compete for resources with technology not only in terms of salaries but also in terms of standards of instrumentation, technical assistance, and so on. In the United States particularly these standards are set by the industrial laboratories and the large government institutes which, for example, generate the demand for commercial instrumentation, and so forth.

Because of its rapid diffusion there is a tendency toward equalization of standards in the international sphere, especially in "big science." As costs grow to the point where a nation cannot produce its "share" of world science in a field, or where the same situation results from the existence of too large a "critical size" in certain fields, pressure will be generated for the creation of international organizations with sufficient concentrations of talent and resources to compete on the world scale. Thus in high-energy physics, the field where this process is most evident, there are only three competitors: the United States, Europe as a whole, and the U.S.S.R. One can now foresee a period of accelerated internationalization of science in field after field, with more and more a world common market of ideas and people. This will happen despite political inhibitions. Unless there is some sort of political or social crisis, the pressures toward internationalization are too strong.

In the more gentlemanly eighteenth and nineteenth centuries science was to such a degree an international activity that scientific communication was permitted to continue across national boundaries even between enemy countries. The scientific enterprise was regarded as truly above and beyond national considerations. However, as technology has come to depend more on science, there has been a tendency for even pure science to enter the arena of national rivalries. This has, perhaps, been enhanced by the fact that national competition has proved the most persistently effective argument in individual countries for increasing national science budgets. Because it is much more tangible than the four general aspects of science I have enumerated, it is easier to use in convincing the public and the politicians, and scientists have found themselves caught between the inherently international ethos of science and the practical necessities of financial support. This should not be held too much against scientists. If international competition has been used as an argument for the support of science, it will not be the first time in history nations have done the right things for the wrong reasons. Politicians are more relaxed with the idea that it is more important to get into power than to take the right position on every single issue. On the other hand, scientists themselves are acutely aware that when the wrong arguments are used too long to support the right actions, the integrity of the actions themselves may be compromised.

GENERAL CONCLUSIONS

In the foregoing paragraphs I have tried to suggest four different ways of looking at the relationship between science and society, and to indicate their implications for science policy. There is some element of validity in each, and each actually puts its emphasis on a somewhat different region of the continuous spectrum between fundamental research and applied technology. The technical overhead theory places greatest emphasis on the technology end of the spectrum while the autonomy and consumption good theories tend to look more at the pure science end. However, science and technology really form a single strongly interacting system. While the purest science may not interact at all with technology and appears as a completely isolated enterprise of a tiny elite, it does interact, but only through a long chain of interrelated activities. This interconnection means that events occurring anywhere in the system have repercussions through the whole system. Most of these are extremely difficult to foresee. Many of the current demands for better scientific planning are probably as naïve as the early demands for economic planning. We are like an untrained person suddenly set down in the cockpit of a jet aircraft, with hundreds of dials and levers in front of us, and little clue as to what lever to pull to steer the machine, though knowing if we push one too strongly the giant aircraft, which tends to fly by itself on an even keel, may go out of control, or respond in the exact opposite way from what we intended. As in the case of economic planning we have to develop a much more sophisticated understanding of how the existing system works before we can control it. Just as economics could not really be applied successfully to policy until it learned to distinguish between "ought" and "is," so must science policy rid itself of a certain reformist or missionary spirit before it can become a tool to successfully influence national and international actions. This same confusion between how we would like the economy to operate and how it actually works was what plagued economics thirty years ago. Today our felt need to plan too far exceeds our understanding of how the system of science and technology really operates. The situation is complicated by what I would refer to as the "demographic transition" in science and technology, at least in the developed countries. Until recently, the growth of pure science, and to a large extent the growth of technology, have been manpower limited. This may still be true in Europe, but it is not so obviously true in the United States. Rather, the limitations there appear to be increasingly in resources and in our ability to mobilize resources. Whether this is a real transition or only a temporary pause in growth for the redeployment of effort is difficult to say. My own belief is that the problems faced by the world are such that only a continuing and growing mobilization of intellectual effort offers any possibility of solving them. Thus I suspect that the system may continue to grow for several years more, but the signs of a transition are still not far away.

FOUR □ The Scientific Adviser

The following essay first appeared in a volume of essays edited by Robert Gilpin and Christopher Wright under the title Scientists and National Policy-Making. *Like most of the essays in this volume it represents the end product of my thinking on a problem which had previously been presented in a number of conferences and university seminars. Thus it benefited greatly from ideas and criticisms contributed by colleagues and audiences. It also reflects much of my own experience and observation as a participant in the advisory process in the federal government. Despite the contributions of others, however, I must take responsibility for the views expressed.*

Much of what has been written about the role of scientists in government during the period after World War II has been the work of political scientists, who used mostly material available in the public literature, especially congressional hearings and newspaper accounts of the statements of prominent scientific figures. One obtains from this literature very little sense of what scientists themselves felt about the process they were engaged in, or what they thought their unique contribution to the political process was. The present essay was an attempt to fill this gap, at least from one point of view. It tends to run counter to much of the recent popular mythology about the role of the scientific adviser, but no doubt substitutes for it some of the mythology of the scientific community itself, and of the participants in the advisory process. At a minimum, however, I think it is a contribution toward a balanced discussion of the problems.

Throughout American history the federal government has used scientific advisory committees made up of part-time outside consultants. Since World War II this practice has flourished, and has even become institutionalized in the form of statutory scientific advisory committees. The function of giving scientific advice to the federal government has begun to assume a professional status, and, as Gilpin points out,[1] a hierarchy of part-time

[1] Robert Gilpin, "Natural Scientists in Policy-Making," in *Scientists and National Policy-Making*, ed. by Robert Gilpin and Christopher Wright (Columbia University Press, New York, 1964), pp. 1–18.

advisory groups has emerged that parallels the bureaucratic hierarchy within the structure of government. This interesting development has accompanied the rapid increase in the use of contracts and grants by federal agencies to support research and development in the private sector—in industry, universities, and research institutes. It is difficult to decide which is cause and which is effect, but there is little doubt that federal support for private research and development and scientific advising have gone hand in hand.

Government scientific advisory committees form a complex interlocking network, and many scientists and engineers are members of committees at several different levels in the hierarchical structure. In some cases this overlap is deliberate; for example, in the Department of Defense the chairmen of the advisory committees to the three military services are automatically members of the Defense Science Board, which advises the Secretary of Defense. In other cases the overlap is accidental; the same individual is co-opted for different committees on the basis of his individual talents and experience. In both instances, this overlap forms a parallel communication network within the federal government which to a very considerable extent circumvents the customary bureaucratic channels. This bypassing of the bureaucracy is probably one of the most important and useful functions of scientific advisory committees. In science and engineering no level of the bureaucracy has a monopoly on new ideas, and the loose nature of the advisory system provides one means by which ideas originating at a low level in the bureaucratic structure can be brought directly to the point of decision without going through regular channels, and new ideas from outside the federal structure can be introduced quickly into governmental operations.

We shall begin with an examination of some functions of scientific advisers, and will then examine the roles of scientists in the advisory process, the qualities and skills of scientists that are called upon when they give advice. Finally, we shall discuss a number of problems and conflicts which arise in scientific advising.

FUNCTIONS OF SCIENTIFIC ADVISORY COMMITTEES

The term "scientific advisory committee" is used generally although such committees are often as concerned with technology and engineering as with science and include many engineers or other applied scientists among their members. The role of the adviser varies greatly, depending upon the level in the federal hierarchy at which his advice is sought and implemented. In general, the lower the level, the more strictly technical the nature of the advice sought, although this is not always the case.

We may distinguish five advisory functions:

1. To analyze the technical aspects of major policy issues and interpret them for policy-makers, frequently with recommendations for decision or action. At the highest levels this often involves the analysis of political issues to determine which issues are political and which can be resolved on a technical or scientific basis. It also involves interpreting the policy implications of technical facts, opinions, or judgments. Familiar examples of this type of advice are such questions as whether to seek a nuclear test ban or whether to resume nuclear testing and, if so, when.
2. To evaluate specific scientific or technological programs for the purpose of aiding budgetary decisions or providing advice on matters affecting public welfare or safety. Many important decisions involving choice between alternate weapons systems, or determinations of whether to proceed with major technological developments such as civilian nuclear power, are of this nature. The review function

of the Advisory Committee on Reactor Safeguards of the Atomic Energy Commission (AEC) is an example of such use of scientists in the area of public safety.

3. To study specific areas of science or technology for the purpose of identifying new opportunities for research or development in the public interest, or of developing coherent national scientific programs. Such studies may be science-oriented, that is, concerned with specific scientific disciplines as in the work of the Committee on Oceanography of the National Academy of Sciences (NAS)[2] or the Panel on High Energy Physics of the President's Science Advisory Committee (PSAC). They may also be need-oriented, that is, concerned with the use of science for a specific social purpose, as in the case of the recent study on natural resources made by the National Academy.[3] These studies may be conducted either on a continuous or *ad hoc* basis.
4. To advise on organizational matters affecting science, or a particular mission of an agency involving the use of science or scientific resources. The continuing advisory boards of the military services and the Defense Science Board serve mainly this function. A recent example is the recommendation of a PSAC panel which led to the establishment of the National Aeronautics and Space Administration (NASA), or the recommendation of another panel for the creation of the Federal Council on Science and Technology (FCST).[4]
5. To advise in the selection of individual research proposals

[2] NAS–NRC, Committee on Oceanography, *Oceanography 1960 to 1970* (Washington, D.C., NAS–NRC, 1959).

[3] NAS–NRC, Committee on National Resources, *Natural Resources; A Summary Report* (Washington, D.C., NAS–NRC, 1962), Publ. No. 1000.

[4] PSAC, *Strengthening American Science* (Washington, D.C., USGPO, 1958).

for support, as in the so-called "study sections" of the National Institutes of Health, or the Advisory Panels of the National Science Foundation.

With the possible exception of the fifth designation, none of these functions is purely scientific in nature or depends purely on technical knowledge or expertise. All recommendations involve nontechnical assumptions or judgments in varying degrees. In some cases the nontechnical premises are provided by the policy-maker seeking the advice, but more often they have to be at least partly supplied by the scientist himself. For instance, in the judgment as to the safety of a nuclear reactor installation, "safety" itself is a relative term. In a sense the only truly safe reactor is the one which is never built. Every technical judgment on safety is actually a subtle balancing of risk against opportunity — the tiny risk of injury to the public against the advantages of nuclear power. Yet the administrator seldom makes clear to the adviser just how this balance between advantage and risk is to be achieved. Much of the apparent disagreement among scientists over the danger of fallout from bomb tests stems not from a conflict as to actual technical facts, but rather from a difference of views as to the relative weight to be assigned, on the one hand, to the political and military risks of test cessation and, on the other hand, to the possible threat to human welfare resulting from the continuation of testing in view of the large uncertainties in our knowledge about radiation effects.

In a somewhat oversimplified way, the functions of the scientific adviser may be divided into those concerned with science in policy and those concerned with policy for science. The first is concerned with matters that are basically political or administrative but are significantly dependent upon technical factors — such as the nuclear test ban, disarmament policy, or the use of science in international relations. The second is concerned with the development of policies for the management and support of the national scientific enterprise and with the selection and evaluation of substantive scientific programs. It is not possible to draw a sharp line between these two aspects. For example, the negotiations over a nuclear test ban — which certainly involved science in policy — led directly to recommendations for a greatly expanded program of federally supported research in seismology, with quite specific suggestions as to the particular areas of promise — which is obviously policy for science. Conversely, the proposal for an International Geophysical Year, which was essentially a very interesting and exciting scientific proposal, involved highly significant political considerations and in many ways became an important tool of U.S. foreign policy.

ROLES OF SCIENTIFIC ADVISERS

What are the particular qualities and skills demanded of scientific advisers? What kind of knowledge and experience do they bring to bear? Although the public image of the scientific adviser is primarily that of expert or specialist, the way in which he is actually used is much broader. One may distinguish at least seven different roles which closely relate to the above five functions.

The scientist or engineer is used for his expert knowledge of particular technical subject matter, as in the study sections of the National Institutes of Health whose function it is to rate research proposals.

He makes use of his general "connoisseurship" of science and scientific ways of thinking. In this role he is required to transfer his scientific experience from fields in which he is expert to fields of science and technology with which he is only generally familiar. He is used for his ability to understand and interpret quickly what other experts say, to formulate general policy questions involving scientific considerations in terms suitable for presentation to a group of experts, and to detect specious or self-serving arguments in the advice of other experts.

He makes use of his wide acquaintance within the scientific community and his knowledge of scientific institutions and their manner of operation. In this role he often helps by suggesting key technical people to serve in full-time government positions, and even in persuading the preferred candidate to accept the appointment. He may also predict the effects of government policies or actions on scientific institutions, and serve the ends of both government and science by defending the scientific community against ill-advised or inappropriate administrative procedures of government affecting the conduct of research and development.

The public administrator often makes use of the confidence and prestige enjoyed by scientists in order to obtain backing for projects or policies which he has already decided to undertake. While this particular use of scientific advisers is not necessarily to be deplored, it is subject to abuse. In many cases the administrator may "stack" his committee to obtain the advice he wants, or to obtain a "whitewash" for doubtful decisions. On the other hand, an advisory committee may often legitimately be used to help an administrator rescind an unwise decision without humiliation or embarrassment. A famous example is the appointment of an advisory committee of the NAS to investigate the National Bureau of Standards at the time of the battery additive controversy.

The scientist is increasingly being used as a specialist in policy research. This practice began during World War II with the development of the science of operations research, mainly by physicists in Great Britain and the United States. It involves the construction of mathematical models of varied military situations and the quantitative prediction of the military results of the use of varied weapons systems and strategies. During the postwar period this methodology was extensively elaborated by mathematicians, economists, and theoretical physicists, and has been institutionalized in such organizations as the RAND Corporation, or the Weapons System Evaluation Group which serves the Joint Chiefs of Staff. A number of amateur and professional groups have also developed to carry on policy research in the field of disarmament or arms control. In all these examples natural and social scientists collaborated in order to bring to policy problems the methods of analyzing problems which are characteristic of the physical sciences.

Scientists are often sought for policy advice merely because the scientific community provides a convenient and efficient process for selecting able and intelligent people. One is reminded of Macaulay's dictum that he wanted "to recruit university graduates in the classics not because they had been studying the classics, but because the classics attracted the best minds, who could adapt themselves to anything."[5] If one substitutes nuclear physics for classics in this quotation, one has a basis for the selection of certain kinds of advisory committees. It is also probably true that physicists have a way of simplifying problems which is especially useful to harassed administrators — a capacity which has its pitfalls, since the temptation to oversimplify is always present.

According to C. P. Snow, science is more oriented toward the future than most other disciplines and scientists are animated by a belief that problems are soluble.[6] Such natural optimism, even when unjustified, is an asset in attacking disarmament problems which have resisted solution for such a long time. It is undoubtedly a characteristic which has brought

[5] Referred to in D. K. Price, "The Scientific Establishment," in *Scientists and National Policy-Making,* ed. by Gilpin and Wright (Columbia University Press, New York, 1964), p. 25.

[6] C. P. Snow, *Science and Government* (Cambridge, Mass., Harvard University Press, 1961). See also W. R. Schilling, "Scientists, Foreign Policy, and Politics," in *Scientists and National Policy-Making,* ed. by Gilpin and Wright (Columbia University Press, New York, 1964), pp. 144–173.

scientists into policy advisory roles even in areas where they are not especially qualified. Similarly, science forms the most truly international culture in our divided world, and scientists probably enjoy better communication with their counterparts throughout the world than members of any other discipline. Consequently, it was natural that scientists should lead in the cultural penetration behind the Iron Curtain, and in organizing and promoting joint international activities and exchanges.

PROBLEMS IN SCIENTIFIC ADVISING

The preceding discussion demonstrates that specific expertise is only a small part of the scientist's role as adviser. It is this very fact that is responsible for much of the controversy surrounding the present role of scientists in government and especially their role in the White House and the Executive Office of the President. The criticism of scientists is epitomized by Harold Laski's stricture on expertise:

> It is one thing to urge the need for expert consultation at every stage in making policy; it is another thing, and a very different thing, to insist that the expert's judgment must be final. For special knowledge and the highly trained mind produce their own limitations which, in the realm of statesmanship, are of decisive importance. Expertise, it may be argued, sacrifices the insight of common sense to intensity of experience. It breeds an inability to accept new views from the very depth of its preoccupation with its own conclusions. It too often fails to see round its subject. It sees its results out of perspective by making them the centre of relevance to which all other results must be related. Too often, also, it lacks humility; and this breeds in its possessors a failure in proportion which makes them fail to see the obvious which is before their very noses. It has also a certain caste spirit about it, so that experts tend to neglect all evidence which does not come from those who belong to their own ranks. Above all, perhaps, and this most urgently where human problems are concerned, the expert fails to see that every judgment he makes not purely factual in nature brings with it a scheme of values which has no special validity about

it. He tends to confuse the importance of his facts with the importance of what he proposes to do about them.[7]

The view so eloquently expressed by Laski is sometimes echoed by political scientists and government administrators in referring to the present influence of scientists in the high councils of government. While it is a valid warning against the uncritical acceptance of scientific advice, particularly where tacit ethical and political judgments are involved, it is not exactly a fair description of the way in which the senior scientific advisers have exercised their responsibilities. Scientists in government have not claimed that their advice should be overriding, but they do insist on the value and importance of this advice in reaching a balanced decision in matters involving the use of scientific results. This is true even when the scientist is speaking primarily as a citizen outside his area of special competence. Especially on matters of military technology, scientists are often in a position to exercise their political and ethical judgments as citizens in a more realistic and balanced manner than other citizens. Precisely because they are so familiar with the technological aspects, they are able to concentrate more on the other issues involved without becoming overawed by mere technical complexities. While scientific advice is not free of bias, or even of special pleading, it is probably more free of prejudice than much other professional advice, and at least has the virtue of providing a fresh perspective unprecedented in government councils.

A number of problems arise in scientific advising which are, in one way or another, related to the type of problem raised by Laski in the above quotation. Discussion of these problems can be organized under eight topical headings: (1) the selection of advisers, (2) the scientific adviser as the representative of

[7] Quoted by I. L. Horowitz, "Arms, Policies, and Games," *American Scholar*, XXXI, No. 1 (January 1962), p. 94. The original source of the quote is Harold Laski, "The Limitations of the Expert," *Fabian Tract* No. 235.

science, (3) communication between the scientist and the policy-maker, (4) the relation between advice and decision, (5) the responsibility of the adviser, (6) the resolution of conflicting viewpoints, (7) executive privilege, and (8) conflict of interest.

SELECTION OF ADVISERS

Although the method of selecting members of a scientific advisory committee depends strongly on the function of the committee, the usual procedures are rather informal and based, for the most part, on personal acquaintance. This is especially true of committees that operate at the higher levels of policy discussion. The final selection is made by the executive to whom the committee is responsible, but usually the executive accepts the suggestions of present members of the committee or of other advisory committees. Thus the advisory role tends to become self-perpetuating, and constitutes a kind of subprofession within the scientific professions. Certainly administrative skills and some degree of political sophistication are factors almost as vital as scientific competence and reputation in the selection of members for the top committees. Experience in one of the major wartime laboratories, especially the M.I.T. Radiation Laboratory and the laboratories of the Manhattan Project, or an apprenticeship on one or more of the military "summer studies," still appears to be a useful qualification for scientific advising. There is as yet little sign of a change of generations that would affect this pattern. Even the relatively few younger scientists who have filtered into the higher-level advisory committees are often students of one of the wartime giants like Rabi, Teller, Oppenheimer, or Fermi. Full-time administrative experience in the federal government or long experience on lower-level advisory committees or panels are also important qualifications. One of the most common methods of evaluating possible candidates for membership on the PSAC is a tryout on one of its numerous specialized panels. In this process of selection for advisory committees the characteristics deplored by Laski often tend to be weeded out.

REPRESENTATION OF THE SCIENTIFIC COMMUNITY

The higher-level committees are often criticized for inadequately representing some particular disciplines, certain kinds of institutions, or some points of view on major national questions such as disarmament. For example, the PSAC has been criticized for having too many physicists and not enough engineers, too many academic scientists and too few industrial scientists, too many scientists from the east and west coasts and too few from the central areas of the nation, too many scientists who are prepared to negotiate with the Soviets and too few representing the school of thought of which Professor Edward Teller is the most articulate spokesman. All of these criticisms have some factual basis, and yet it is essential to remember that scientific advisory committees are not legislative bodies, that the ability to reach a large measure of consensus and settle matters by a good deal of give and take in rational argument is much more important to the policy-maker than assurance of equal representation for all the "estates" of science and technology. People with very strong viewpoints which are impervious to rational argument or compromise merely tend to lead to a hung jury which does not help the decision-maker. A majority vote is much less useful than a well-reasoned consensus in providing scientific advice.

Many of the criticisms regarding lack of representation are either untrue or grossly exaggerated. For instance, there have always been industrial scientists on the PSAC. Although a majority of the Committee are physicists, it also consists of engineers, life scientists, and members of the medical profession. A high representation of academic scientists on the presidential committees and panels is balanced by a predominant representation of industrial scientists and research directors on the Defense Science Board and the top advisory committees of the military services. Yet to some extent this is beside the point. The members of the Committee are supposed to be selected for their ability to look at problems in broader terms than those of their own corner of science. The advice of a committee is not the sum of the individual expertise of its members, but a synthesis of viewpoints of people accustomed to looking at problems in the broadest terms. The high proportion of physicists stems from their wartime experience and their subsequent military advisory experience, for it must be remembered that scientists came into the top advisory role in government via their contributions in the national security field, and it was only later that they became concerned with the broader problems of basic science policy and the impact of science on international affairs.

It is important to bear in mind that on any given problem it is the practice of most scientific advisory committees to delegate much of the groundwork to panels whose memberships are carefully chosen to reflect the scientific and engineering skills required in the solution of particular problems. Furthermore, if the issue to be resolved is politically controversial, a special effort is made to ensure representation of a wide spectrum of viewpoints on the panel even though such a variety may not be represented on the parent committee. At the same time it is important to avoid people who are so committed to one view that a discussion of real significance is impossible. A surprising measure of agreement can be reached by a group of scientists of divergent views when they are partially protected by individual anonymity and not constrained by the need to be consistent with previously voiced public positions.

The Special Assistant to the President for Science and Technology, as well as the PSAC, is often thought of as the "spokesman" of science. It is sometimes said by both scientists and nonscientists that the members of the PSAC should be lobbyists for the interests of the scientific community and promoters of science. This feeling came about partly because the creation of the Office of Special Assistant was the result of public concern over inadequate national attention to the cultivation of national scientific strength. An important part of the task initially facing the Special Assistant was to promote the support of science, particularly basic science, in every way possible.[8] Actually, the PSAC and the Special Assistant are not, and should not be, official spokesmen for science but are organs of government. They regard themselves not as advocates of science, but as mediators between science and government. In this function they feel an obligation to take into account the needs and interests of the government as a whole and not just the needs of the scientific community. It is the NAS and its committees which are and should be the advocates of science. The enthusiasts for a particular field of science are represented on these committees, but their reports are criticized and reviewed by the Special Assistant, by the PSAC and its special panels, and by panels of the FCST. This process is an attempt to balance the demands of a special field against the overall requirements of science, and to adjust the requirements of a special program to the fiscal and administrative limitations of the federal government. It is still an imperfect process which has not yet been fully tested. In recent years science budgets have been rising so rapidly that the problems of maintaining balance between fields of science within the confines of severely limited resources have not had to be faced.

COMMUNICATION WITH THE POLICY-MAKER

Scientific advisers are frequently criticized for their tendency to expand their role beyond that of purely technical advice into

[8] R. N. Kreidler, "The President's Science Advisers and National Science Policy," in *Scientists and National Policy-Making,* ed. by Gilpin and Wright (Columbia University Press, New York, 1964), pp. 113–143.

the political, financial, and organizational spheres. The point is a valid one, but this tendency is inherent in the nature of scientific advising rather than a deliberate effort at usurpation on the part of scientists. Science makes progress largely by redefining the key questions and problems. The scientist approaching a policy problem wants to begin by understanding the whole problem, in order that he can break it down into its components in his own way. This habit, learned from his experience as a scientist, has been strongly reinforced by his wartime experience in solving military-technical problems. As Rabi puts it: "[T]he military man who doesn't come clean on the *whole* problem is like a patient who doesn't tell his doctor all the symptoms." An important part of the task of a scientific adviser is to define just what the technical issues are that have a bearing on a given policy decsion. The policy-maker who tries to define the technical issues himself will not obtain the best advice from his scientific advisers. Similarly, the scientific adviser has an obligation to interpret his advice in terms of its policy implications, while at the same time trying to make explicit the nontechnical assumptions which necessarily underlie his recommendations. The policy-maker and the adviser together have the obligation to differentiate between technical and political questions, as well as between what is actually known and what is a matter of professional judgment.

The science adviser cannot always be blamed when he steps outside the technical area. Too often the politician or administrator is tempted to throw the onus of difficult or controversial political decisions onto his scientific advisers. An important decision may be much more palatable to the public and to Congress if it is made to appear to have been taken on technical grounds. In the early days of the negotiations on a nuclear test ban, both scientists and diplomats fell into the trap of believing that the basic issues were primarily technical ones which could be resolved by discussions among experts, if not at the time, then later on as new scientific knowledge became available. Subsequently, it became increasingly clear that the really difficult issues were related to the degree of assurance which the United States felt it must have against the conduct of clandestine underground tests, and to the Soviet judgments as to what would be the acceptable degree of penetration of their military security. The importance of detecting clandestine underground tests has been differently estimated by the United States, depending on judgment as to the military decisiveness of tactical nuclear weapons. The winds of public and governmental opinion appear to be too much influenced by day-to-day changes in technical developments and ideas such as the "Latter hole" or the Tamm "black boxes." [9]

The nonscientist often has an exaggerated faith in the exactness of physical science, and has great difficulty in distinguishing between what is known with a high degree of certainty and what is only a matter of reasonable probability or scientific hunch. Under pressure to make concrete recommendations, the scientist has often tended to exaggerate the validity of his data and to permit the administrator to erect an elaborate superstructure of policy on a very flimsy technical base. Something of this sort happened twice in the nuclear test ban negotiations when much too general conclusions were drawn from fragmentary data obtained from one particular U.S. underground test. As in military decisions, policy decisions can seldom be made with all the necessary information available, and the scientist who refuses to commit himself until he considers his data completely adequate is not very useful to the administrator. He does, however, have an obligation to explain the areas of uncertainty to his political master and to prepare him as best he can for technical surprises.

[9] U.S. Disarmament Agency, State Department, *Geneva Conference on the Discontinuance of Nuclear Weapons Tests; History and Analysis of Negotiations* (Washington, D.C., USGPO, 1961), Publ. No. 7258.

ADVICE VERSUS DECISION

Political scientists writing about the PSAC and other advisory groups have often tended to confuse advice with decision, thereby investing the PSAC with a power and responsibility which it does not in fact possess. It is true that the scientist's public prestige occasionally gives his advice an influence and authority with decision-makers. He thus has power which is in some respects equivalent to, though not identical with, political power. Ultimately the scientific adviser recognizes that his influence in government rests solely on the degree to which his views are verified by subsequent events. He has no true constituency to give his advice political force, and perhaps to a greater degree than in the case of any other type of adviser his influence must rest with the persuasiveness of his arguments. Professor Bethe's prestige as an adviser was, perhaps unjustifiably, dimmed by his failure to anticipate the possibility of decoupling in underground nuclear explosions. At the same time, his prestige would have suffered much more if he had not been so quick to accept the new technical suggestion and examine it on its own merits without reference to its effect on his own deeply held political convictions regarding the desirability of a test ban.

Congressman Melvin Price has publicly attacked the PSAC for its role in the decision to abandon the aircraft nuclear propulsion program (ANP),[10] implying that it alone was responsible for this decision and that its recommendation was based on political and budgetary grounds. While there is no doubt that nontechnical considerations played an important role in the actual decision, the PSAC's part was to give advice on primarily technical grounds, as has been clearly stated by

[10] Melvin Price, "Atomic Science and Government — U.S. Variety," an address delivered to the American Nuclear Society in Washington, D.C., on June 14, 1961.

Kistiakowsky.[11] Contrary to Congressman Price's implications, the responsibility for the ANP decision was shared by many administrators and advisers; the voice of the PSAC was only one among many voices, and this voice was probably not decisive. Such decisions within the executive branch are seldom reached through the advice of a single group or individual but are the result of a gradually evolving consensus among many advisers.

THE PROBLEM OF RESPONSIBILITY

The scientific adviser differs from the military adviser in government in that he is seldom responsible for carrying out his own advice, or even for the consequences of the advice he has given. This situation has both advantages and disadvantages. It can lead to the particular type of irresponsibility which Laski describes above so graphically, or it can permit a degree of detachment and objectivity which would be difficult to achieve if the adviser were more deeply concerned with the consequences of his advice. The advice given by the Joint Chiefs of Staff has always been plagued by parochial service interests, for example.

The problem of the responsibility of advice comes to the fore in the field of budgetary decisions affecting science. Unfortunately, advisory committees of scientists are seldom presented with the hard choices between attractive alternatives which usually concern the budgetary officer or administrator. The competing claims of different fields of science have yet to be squarely presented to a scientific advisory committee. When confronted with the virtually unlimited opportunities in a scientific field, the advisory committee is tempted to recommend expansion without much reference to other alternatives. The

[11] George B. Kistiakowsky, "Personal Thoughts on Research in the United States," in *Proceedings of a Conference on Academic and Industrial Basic Research* (Washington, D.C., USGPO, 1960), NSF 61–39, pp. 49–53.

balance between scientific fields in the past has been determined by the somewhat accidental resultant of many pressures, both political and scientific. The committees that recommend expansion, while not achieving all they hoped for, generally see enough effect from their recommendations to be reasonably satisfied. In an era of expanding scientific budgets this has worked fairly successfully, and the general balance of scientific effort seems to have been preserved. The scientific adviser may find himself faced with an entirely different responsibility when the time comes, as it inevitably will, that scientific budgets level off. The scientist who has to live with his professional colleagues outside government, especially in universities, will find himself torn between his natural inclination to appear as the champion and promoter of science on every occasion and his sense of responsibility as a government adviser.

RESOLUTION OF CONFLICT

In controversial issues the ideal advisory committee is one which succeeds in enlarging the area of agreement and reaching as wide a consensus as possible. A wise committee or panel can often succeed in narrowing the disagreements on a complex issue to a few technical issues which might be resolved by further research or to a clean-cut set of political alternatives which can then be resolved by the administrator. The committee which strives for consensus at all costs usually ends up with a series of pious platitudes which are useless to the policymaker — and useless in a peculiarly irritating and frustrating way. Having reached the widest possible area of agreement the committee should then attempt to formulate the disagreements as clearly and objectively as possible. Recommendations should generally be formulated in terms of forecasts of the probable consequences of alternative actions rather than in terms of exhortation. The responsibility for these tasks rests, for the most part, with the chairman, on whom usually falls the further duty of interpreting both the agreements and the disagreements to the policy-maker. This calls for unusual objectivity and detachment on the part of the chairman. In some cases it is wise for the policy-maker himself to hear the arguments of both sides directly from the proponents rather than filtered through the chairman.

It has sometimes been suggested that major issues involving technical advice should be resolved by a sort of adversary procedure, as in a court of law. There are instances when this is desirable, as, for example, when the rights or interests of individuals or groups may be in jeopardy as a result of the decision to be made. However, on broader policy issues the advisory process should be designed to encourage convergence rather than divergence of views. An adversary procedure tends to produce a polarization of viewpoints which then must be resolved by the policy-maker himself. The administrator would be forced to immerse himself in the technical details of every decision as does a judge in patent litigation. Given the many decisions which have to be made by every administrator, and especially by the President, such a process would be ludicrously cumbersome and would paralyze decision-making.

The alternative suggestion that every advisory committee should include a devil's advocate is probably a good one. There are times when the chairman should deliberately assume this role. Sometimes it is the only way that the strongest arguments for the committee's final position can be brought out and potential objections to its recommendations anticipated.

On the whole the greatest occupational hazard of advisory committees is not conflict but platitudinous consensus.

EXECUTIVE PRIVILEGE

No aspect of the PSAC has received as much criticism as the sheltering of its deliberations and recommendations under executive privilege. Other scientific advisory committees enjoy varying degrees of privilege, but never to the extent of those operating directly under the aegis of the White House. Of course, this is only a part of a more general source of irritation between the executive and legislative branches. The irritation has been directed at the PSAC only because its advice has frequently been followed, and has sometimes been against important agencies or congressional positions and projects. In practice, Congress, and for that matter, the public, have seldom been denied access to the technical and nontechnical considerations on which the PSAC advice was based; only questions of who said what are withheld. In some instances the PSAC panels have actually been reconstituted as panels of other agencies in order that their views could be made public without revealing the advice precisely as it was given to the President. In the early days of the PSAC, the Special Assistant, Dr. Killian, frequently advised the heads of agencies about the recommendations he intended to make to the President, and this has remained a common practice.[12] The decision to drop the ANP project, for example, was actually concurred in by the Director of Defense Research and Engineering and a panel of the General Advisory Committee of the AEC. Most of the technical and military considerations involved in that decision were covered in testimony before Congress by Dr. Herbert York and others.

There is a general feeling in the scientific community that J. Robert Oppenheimer was persecuted because of the unpopularity of his advice, supposedly given in private. The protection of executive privilege is sometimes necessary to induce scientists to join panels. They feel, rightly or wrongly, that their private interests may be vulnerable to reprisal by Congress or by powerful agencies which may be adversely affected by their recommendations.

A different type of problem has arisen when scientific ad-

[12] See Gilpin essay in reference 1.

visers have chosen to speak out publicly on issues with which they have also been concerned as scientific advisers under the mantle of executive privilege. People opposed to their views feel that advisers who take public stands on controversial issues are trying to have their cake and eat it too. These advisers have, however, been willing and even eager to testify before Congress, with the exception of specifics like who recommended what to the President. The suspicion may remain that their testimony is filtered or distorted by the omission of privileged matter, but it is hard to see how this problem is any different for a presidential adviser than for an agency or department head who also gives privileged advice to the President. Perhaps the principal difference is that the part-time adviser does not consider himself under political or administrative discipline, and so may feel more free to volunteer a public statement at variance with the official line of the moment. The administrator who differs from the official line is constrained from expressing his difference except when called upon to do so while testifying under oath.

This is a complex question, but its facets are not peculiar to scientific advisers, except insofar as scientists may enjoy greater public prestige than other experts. Because of their relative newness in the higher government councils, the reputation of scientists is less tarnished by special pleading or self-serving, and it is hoped that in their public statements scientific advisers will bear in mind their responsibility to preserve the reputation for objectivity which scientists generally enjoy, and which is their greatest asset in the political arena.

Speaking out on public issues should not in itself be considered an abuse of executive privilege. Under the Eisenhower Administration, the Special Assistant often felt hampered by the rigidity of the practice of executive privilege, which is even more enforced when the Executive and Congress are controlled by different parties. There were times when the Special Assistant was unable to testify although it would have been to the interest of the government for him to do so. Reorganization Plan No. 2, which became effective in June 1962, created the Office of Science and Technology and gave the Special Assistant two hats — one as a confidential White House adviser and the other as statutory Director of the Office, subject to Senate confirmation. One of the purposes of providing such statutory underpinning to the science advisory role was to permit the Director to testify before Congress and thereby formally defend Administration positions on new science legislation, on budgetary matters affecting basic science, and on the coordination of federal scientific programs. As a result of this reorganization, the area which we have called "policy for science" can become the subject for congressional testimony, while the area which we have called "science in policy" remained under executive privilege.

In the reorganization, the PSAC remained in the White House and continued to enjoy the protection of executive privilege. It is possible that if all the PSAC members had been made subject to Senate confirmation, this might ultimately have led to political control of science and to a serious threat to the independent and apolitical nature of the Committee. This nonpartisan character was explicit in the PSAC's original charter from President Eisenhower, and was recognized by President Kennedy through his continuation of the membership after the change in administration. Offsetting the fear of partisan control is the historical fact that there has so far been no problem of partisan politics with respect to the National Science Board, whose membership is subject to Senate confirmation.

A possibly more serious problem resulting from the new status of the Special Assistant is the amount of time involved in congressional relations that must now be added to that required for all the other responsibilities of the office. A major part of the past effectiveness of the whole presidential science advisory operation has been due to its compactness and lack of bureaucratization. The Special Assistant and the PSAC were able to be highly selective in the problems undertaken for study. With greater congressional visibility the Special Assistant may find himself forced to know less and less about more and more, and to depend increasingly on staff work rather than first-hand study for his expression of views. Since he has no decision-making or executive responsibility, other than for his own small staff, he can, I believe, avoid this dilemma by proper selectivity with respect to the type of things on which he chooses to testify. The Director of the Bureau of the Budget is in an analogous situation, and has so far lived with it successfully by delimiting his areas of testimony.

The preservation of executive privilege is clearly essential if the President is to get honest and independent advice. Too frequently the violently unpopular position of today becomes government policy tomorrow. If the adviser could not look ahead without fear of political crucifixion, policy would soon become frozen in a static mold. Furthermore, it would be ossified at the very moment of generation, when it should be most fluid and dynamic.

CONFLICT OF INTEREST

Nearly two thirds of all the nation's research and development is financed by the federal government, and 55 percent of the work is carried out through contracts and grants from the government to the private sector. Much of this work is of a nature that is unique to government financing. It is usually impossible to recruit advisers in these fields unless they are closely associated with institutions whose work is heavily financed by the federal government. In its modern guise, the problem of conflict of interest is much more subtle and complicated than envisioned in the conflict of interest statutes, which were designed originally to prevent government officials from directly and personally benefiting from their official position.

Until recently the law was not very clear on the extent to which it was applicable to part-time advisers. By a strict interpretation, most of the complex advisory structure in the federal agencies would have been made illegal. The best that could be done from a practical standpoint was to avoid situations in which government consultants were in positions to affect directly a specific contract or grant with an institution in which they had a substantial interest, or to which they owed an allegiance. Many government consultants, and especially scientific advisers, have acted on the assumption that the statutes applied only to employees or stockholders of profit-making institutions, arguing that as long as a government contract could not affect their personal compensation or financial status there was no possibility of conflict or even impropriety. Few advisers were more than vaguely aware of the conflict of interest statutes; they could be excused for not taking them very seriously since there had been no prosecution under these statutes against a part-time adviser, and therefore the law had never been tested or interpreted in the courts. Most scientists feel that the best protection for the government lies in full and public disclosure of all the outside interests of the advisers together with sufficient technical competence and general sophistication on the part of the administrator receiving the advice to enable him to make allowance for possible bias and discount it in his own mind. In many instances the loyalties of a scientist or engineer are much more closely identified with his professional community than with the institution which pays his salary. Most scientific advisers would assert their ability to be objective even when the interests of their own institution are involved, though they may recognize the undesirability from the standpoint of public relations of having a potential conflict of interest.

Recent legislation and legal interpretation have clarified the concept of conflict of interest, especially as it pertains to part-time consultants. While this legislation does not fully embody the principles enumerated above, it does permit the government to retain the services of its advisers, and to operate with advisory boards and panels much as in the past, though with greater circumspection as to the public record.

No government adviser can be free of bias. Although no statute recognizes the possibility of a conflict of interest within government, every administrator is well aware of the agency biases harbored by federal employees who give scientific advice. The scientists of the three military services tend to favor strategies and weapons systems which lead to the aggrandizement of their service, and the professional military scientist can seldom be depended on for unbiased advice on disarmament.

In fact it is often to counteract the effects of conflicts of interest within government that administrators have sought advice from the private scientific community. This community, in turn, has its own bias toward contracting out as much research as possible rather than doing it within government. Government scientists generally favor government laboratories and are opposed to the practice of contracting out either research or development. University scientists have a bias favoring basic research. These biases are seldom consciously self-serving; they are merely a part of human nature. "What's good for General Motors is good for the country" is not a sentiment unique to a famous Secretary of Defense; there is a little bit of it in every institution and profession. Some advisers are more able than others to recognize this sentiment in themselves and consciously strive to discount it in formulating their advice. These are generally the people that make good advisers. There is also an opposite danger of leaning over backwards, of failing to defend the interests and values of science because of a fear of the appearance of pleading the cause of a special interest.

Issues of the kind described above are much more important

in scientific advising than are the obvious conflicts of interest involving personal profit. Yet it is basically impossible to resolve such issues by any legislative prescription. As long as the federal government retains the necessary competence within its own full-time administrative and technical staff, it has less to fear from advice subtly or overtly biased by self-interest than it does from loss of the benefit of a perspective from outside government. The answer to the conflict of interest problem in scientific advice is not restrictive and negative legislation but positive legislation and administrative action to improve working conditions within government so that it can continue to attract people capable of properly using outside scientific advice. With respect to scientific advisers acting at the higher policy levels, it is important to develop a professional code of ethics connected with the advisory function. Fortunately, among basic scientists there is a stern ethical code associated with the question of scientific credit and priority which is powerfully sanctioned by public opinion within the scientific community. It provides a model for a similarly sanctioned, though unwritten, code with respect to the objectivity of scientific advice.

CONCLUSION

The reader may conclude that this essay has presented an unduly rosy picture of the beneficial effect of scientific advisers in government, and particularly of the operation and influence of the PSAC. There are certainly many critics, especially among those who have disagreed with the influence exerted by the PSAC in the field of disarmament and weapons policy, who would emphatically and sincerely disagree concerning the beneficial effects. Only history will tell who was right with respect to policy.

A major criticism would be of the tendency toward self-perpetuation among the most influential committees, and especially the consequent preservation through two Administrations of a single viewpoint on many questions of policy. Furthermore, critics would argue, the presidential advisers have, through their influence on major appointments, gradually imposed their policy viewpoints throughout the upper levels of the scientific agencies and in the Bureau of the Budget. Although there may be superficial evidence for such an analysis, it will scarcely stand historical examination. The diversity of viewpoints and institutions represented on high-level advisory committees has been much broader than the critics claim. If it can be said that any policy viewpoints have become dominant in government, this has been imposed more by the logic of events than by any particular group of advisers. The advisers merely foreshadowed what would probably have been brought about by events anyway: the creation of an invulnerable retaliatory force; the inadequacy of the policy of massive retaliation; the importance of limited and guerrilla warfare and of conventional arms; the creation of the NASA, the Arms Control and Disarmament Agency, and a research and development section in the foreign-aid program; the centralization of research and development responsibility in the Department of Defense; increased support for basic research and graduate education; the fostering of international scientific activities and of scientific exchanges with the Soviet bloc; the creation of the FCST and its interagency committees; and the creation of the Office of Science and Technology by reorganization plan.

In matters of armament policy the critics have had their day in court and indeed have spoken from a much more firm base of power and influence than the various scientific advisers. Perhaps where the advisers have won out they have done so because their case was more persuasive, rather than because of illegitimate pressures or some sort of inside track to policy.

The scientific community and its influence in government are in a politically exposed position if only because of the magnitude of the scientific enterprise and the success of scientists, especially university scientists, in influencing policy despite their lack of a true constituency or base of power. To an increasing degree the country is looking to the Special Assistant and his supporting advisory staff, and, rightly or wrongly, there will be a growing tendency for Congress to blame his office for everything that goes wrong in the manifold scientific activities of the government. Whether the office can maintain its ability to concentrate on the key issues and not become lost in a rash of bureaucratic brush fires is the crucial question for its future success. It seems probable that the test of the political viability of the present network of advisory committees, and particularly of the presidential advisory structure, will come during the next few years, when budgetary ceilings may put a greater strain than ever before on the advisory committee's task of judging, and when the accountability of scientists to Congress and to the public will be more carefully probed than it has been in the past.

FIVE □ Basic Science and Agency Missions

In May of 1966, Dr. F. Joachim Weyl, then Chief Scientist of the Office of Naval Research (ONR), organized a symposium in honor of the twentieth anniversary of the establishment of the Office of Naval Research. I was invited to prepare one of the papers for this symposium, and the following chapter is the resulting paper, essentially as it appeared in the volume Research in the Service of National Purpose, *published in 1966. The paper drew heavily on specific examples provided me by the scientific staff of ONR. Furthermore, some of the expressions of philosophy drew heavily on memoranda prepared by Dr. Weyl and by Dr. Lewis Branscomb in connection with their participation in a summer study on research support conducted at Woods Hole in the summer of 1965 under the auspices of the Office of Science and Technology. I am especially indebted to Doctors Weyl and Branscomb for their major contribution to this paper.*

The paper addresses itself to the general question of why a mission-oriented organization such as the Navy should support essentially undirected university research and what kinds of research are appropriate for support.

Today we are celebrating the twentieth anniversary of a decision historic for American science, the decision of the Navy to assume leadership in the stimulation and support of basic and applied science in the United States. I do not know how much the original participants in this decision foresaw the significance of what they were doing, or what great consequences it would have for American leadership in world science and technology. I rather think they did appreciate their role. One senses in talking to these early participants in ONR a little of the atmosphere of early Christians spreading a new gospel of the partnership of science and government.

As with all evangelical movements, however, the very success of this one has brought its own problems. We find ourselves today in a period of stocktaking, a time of pause and a time of soul-searching. Perhaps, for the first time since the war, the assumptions on which our science policy for the past twenty years have been based are being seriously questioned and even challenged. There have been several such pauses in the past, but I do not think any in which the fundamentals of our policy have been as profoundly questioned as today.

In some ways this questioning is natural and healthy. After all the original circumstances and environment in which ONR began have been deeply altered by the very successes and imitations which it generated. Today we have at least eight separate federal agencies which are deeply involved in the support of academic research, and many others involved at least peripherally. When it began, the Office of Naval Research was almost singlehandedly responsible for the health and strength of basic science in the United States. It could defend its research investments in terms of the broad Navy dependence on all aspects of modern science with the full knowledge that if it failed to act, the total American scientific strength on which its military strength depended would languish and falter. Today, while many of the same arguments apply, they no longer apply so uniquely, and hence with so much force. It is harder to defend oneself against those who argue, Let NSF do it, or Let ARPA do it. The penalty for failing to support basic science no longer seems so apparent or pressing as it did to the early missionaries, or, what is more important, to their budgetary masters. Perhaps the time has arrived for a reassertion of the role of mission-oriented agencies in the support of basic research, and for a more sophisticated statement of this role in a political environment which is far more knowledgeable about science and technology than was the world of the late 1940s.

It is, of course, easy to say that the mission-oriented agencies should support the basic research that is relevant to their missions, even the training of skilled people who might later serve their mission. This is a statement with which few would now disagree, but it really begs the question. The whole argument is about what basic research really *is* relevant to a mission, and what time horizon one should be talking about. To some extent we seem to be coming into an era where basic research is everybody's business, and therefore nobody's business, except possibly that of the National Science Foundation. Where does the line between "pure" basic research and mission-oriented basic research really lie? Does the proliferation of other research-supporting agencies, and especially the growth of an agency with an explicit mission to support science in terms of its own internal system of values (namely NSF), imply that mission relevance should be interpreted in ever more narrow and specific terms? I do not believe so, whether the matter be judged from the standpoint of the health of science itself, or from the standpoint of the vitality and success of agency missions which depend upon science. I do not believe that science can be divided up into neat little packages, each of which can be related uniquely and bodily, as it were, to the mission of one agency. Two scientific investigations which begin by asking the same questions and using the same methods may end up at an entirely different point merely because of the environment and the communication network within which they are conducted. One should look at mission relevance not in terms of mutually exclusive compartments, but rather in terms of distribution functions with different centers of gravity but substantial overlap in the tails. It is precisely this overlap which provides the internal communication within science which is necessary for the rapid application of science.

The conventional wisdom deals often with the debt of technology to science, but speaks less frequently of the debt of science to technology. Some of the most challenging and fundamental problems of solid-state physics or molecular physics have arisen from studies which were originally suggested by technological needs. Scientific work involves a multiplicity of choices of direction, many of which depend on very small influences in the mind of the investigator. Even in a system of complete scientific freedom the cumulative effect of the small biases placed in the mind of the investigator by his sponsor can have a profound effect on the direction and impact of his research. The mere need to defend what he is doing to a particular sponsor may be the factor which will trigger an important application. It seems to me no accident that both versions of the maser and the laser were conceived in university laboratories devoted to the broad advancement of electronic communications, and sponsored by the military services.

But the invention of these devices, and the race to develop and improve them, also opened up a host of new fundamental questions regarding electric and magnetic interactions in crystals, some of which led far afield from the original device applications. In fact it is striking that each solid-state-device invention has opened up a new branch of pure solid-state research, whose existence was scarcely suspected until the appearance of the device formed the entering wedge into the field and also provided the motivation for its generous support and exploitation.

Our system of science support in the United States is what I like to term a mixed-decision system, and I think one of the major sources of our scientific and technological strength. By a mixed system I mean one in which scientific choices are governed by a wide diversity of priorities, institutional environments, and motivations. Not all pure research is too pure, and not all applied research is too applied. In the apt terminology of Alvin Weinberg, our choices are seldom dominated exclusively by either scientific criteria or social criteria, and our variety of research institutions and sponsoring agencies emphasize these two types of criteria in varying mixes. We recognize the fact that the sponsor of research and its performer may quite legitimately have different sets of motivations and priorities, and that the very tension between the two may be creative.

One of the paradoxes of this age of specialization is that it is harder and harder to delimit the boundaries between scientific fields or the relation between the scientific fields and federal missions. The "modern" technological agencies — Defense, Space, Atomic Energy — draw scientific sustenance from almost all areas of science. Within our present diversity of support systems, how is an agency to decide just what is relevant to its mission? How is it to decide when it can safely depend on the research sponsored by other agencies? When should it depend on inhouse capability, and when should it look to the academic community for intellectual support? When is an activity which it sponsors to be considered above critical size? It would be wrong to expect that these questions could be answered *a priori* — that is, in the absence of past history. What an agency sponsors will depend not only on its own mix of skills and its appraisal of the skills of other agencies, but also on previous history. One cannot set down a system of rules.

Why should a mission-oriented agency sponsor or conduct basic research? It needs a fund of knowledge adequate to the fulfillment of its mission at a satisfactory rate of progress, and to provide it with as broad a range of future technological choices as possible. This fund of knowledge must be effectively available to the technical and managerial arms of the agency, and may have to be adapted to its unique technological requirements. The needed research makes up a chain — or, more precisely a network — extending from work of obvious immediate relevance to the mission back through research in fields of less and less obvious relevance. This chain must be followed reasonably far back toward the most fundamental or abstract fields in order to evaluate how much further the state of the art could be pressed quickly if needed, to appraise the reliability of technical judgments and evaluation of systems based on current understanding, to predict what technical choices may be open to us in the future and what accomplishments might be possible for other nations — in short, to avoid technological surprise. These requirements involve a subtle blending of scientific knowledge and sophistication with knowledge of agency needs and technological thinking. One cannot depend upon the program officers of other agencies to be aware of all the needs of the Navy!

The fund of basic knowledge required by any agency may be divided into three general classes:

1. Fields of science in which the mission orientation admits of no clear limits to agency interest, and requirements differ in both kind and volume from those of any other component of the nation's technological community. In the Navy, underwater acoustics, physical oceanography, and deep-sea technology are clear examples. In such cases the rate of progress which would result from free play of academic interest in the science for its own sake would be unlikely to satisfy the objectives of the agency.
2. Fields of science which are of vital importance to the agency mission, but whose importance is shared almost equally with other agencies. In the case of the Navy, examples which may be cited include electronics, materials, meteorology, and human factors engineering. On the other hand, there may be specific aspects of these broader fields which are of unique relevance to one agency, or have a special flavor which is characteristic of the agency's needs; for example, high-strength metals and composites for deep submergence vehicles in the Navy.
3. Fields which at present show no obvious promise as sources of concepts or research results for near-term agency exploitation but which, in the mainstream of imaginatively advancing science, can produce results of potentially significant repercussions. These areas, such as pure mathematics or elementary-particle physics, may have significance either through the chain of evolution of scientific ideas

which links the fields of greatest scientific interest to those of extensive value, or through the development of new intellectual and experimental tools which are incidental to the science but of wide applicability.

In considering the relevance of a field to agency missions an aspect often forgotten is the significance of these new *tools* of basic research for future technology. Thus, for example, one might have questioned the relevance of the results of nuclear physics to the Navy's mission; yet it is clear that this field, by challenging the most advanced electrical technology, has been an instrument of many technological advances of great importance to the Navy and to Defense; high-powered klystrons, high-speed circuitry, developments in computer software and pattern recognition, high-vacuum technology, and many other fields, of which I will give more detailed examples later.

The approaches to these different types of basic research may be different, but some participation in all three is needed if the agency is to realize the maximum and most timely advantage from ongoing science.

The first category I mentioned — the fields uniquely relevant to the mission of a particular agency — often include areas which do not receive much attention through the internal dynamics of science. In a field such as this, of which underwater acoustics is a good example in the Navy — or indeed, classical hydrodynamics — the agency's support should be limited only by available manpower and promising scientific opportunities. Almost any advance on a broad front will be virtually certain to benefit the agency. In such unique fields, also, a fair fraction of the fundamental knowledge needed will have to be obtained through inhouse or captive laboratory effort. The efficient execution of any applied mission will almost certainly require organized and directed effort in areas where understanding is so limited that basic research is necessary to advance the state of the art. Even here, however, the agency cannot afford to be too parochial, and must be prepared to support some basic research whose immediate relevance may not be apparent. A classic example is ONR's extramural support for fundamental research on long-range acoustic propagation in the 1940s at the Lamont Geological Observatory at Columbia. At the time there was no specific end item in mind, yet when missiles began to look like potential weapons systems, the need arose for test ranges where missiles could impact at sea and be accurately located. The results of the ONR-sponsored research immediately became the basis of the missile impact location system now used at all the test ranges. The discovery of the deep sound channel, now important for long-range underwater detection, was also a direct outgrowth of ONR-sponsored research in one of the oceanography laboratories.

An important function of an agency is to stimulate interest within the broad scientific community on problems important to its future. One way in which this has been done is by finding ways to translate applied problems of an agency into generic scientific problems which can attract the continuing interest and attention of first-rate scientists. To achieve this in any lasting way it is not sufficient to call attention to the importance of a field; it is essential also to demonstrate its intellectual challenge and the opportunity for genuine scientific progress. ONR in its past has shown a particular talent for this kind of imaginative stimulation. Examples are its fostering of basic work in certain key university centers on coastal geography and also on the general theory of port logistics. This work has paid off handsomely not only for the Navy but the DOD generally in the present situation in the Far East. In such areas it is necessary to recognize the significance of a field long *before* its practical importance. Without this kind of imagination, applied research often becomes little more than research on how to solve the problems which will have disappeared or changed unrecognizably by the time solutions are found. The secret of successful research planning is to have the science ready when the time is right.

In fields of research whose importance to a mission is shared with other agencies, the problem is often not so much one of stimulation as of maintaining close contact with a broad area of science and technology and being in a position to influence its evolution and identify the technological opportunities emerging from it at an early stage. Here a combination of inhouse and extramural effort is desirable, with organizational devices to maintain contact with as broad a base of extramural effort as possible. The Joint Services Electronics Program, supported jointly by the three military services, is a good example of the kind of device which is valuable. From the university-based JSE laboratories has flowed a steady stream of contributions, direct and indirect, to the development of naval equipment and weapons systems.

It was from work supported in these laboratories that the ideas for the maser and the laser emerged, for example. Increasing sophistication of signal detection methods, in guidance and control technology, in antenna design, in radio navigation systems, in high-power microwave tubes, and in a host of other areas, has emerged from this long-supported activity.

The third area of support I mentioned earlier is that of fields whose relevance to an agency's mission is not at all obvious or readily demonstrable in a before-the-fact sense. This is, perhaps, the area in which the general philosophy of mission-oriented support of research is most under attack today, and where the attitude of "Let George do it" is most evident both inside and outside the agency concerned. I am not here suggesting that an agency can or should ignore what is happening in other parts of the government, but neither can it afford in the long run to lose contact with the field. Unlike the first two areas I have mentioned, an agency in this area has little concern with the total volume or rate of research. It is not here concerned with stimulating fields which are in danger of falling into scientific backwaters or with supporting a volume of activity consistent with the advancing requirements of its own technology. Rather its involvement might be described as a "listening-post" commitment, a kind of scientific early-warning device. This listening-post activity can often be best achieved through extramural support. The relevance of the work is usually not sufficiently certain to justify the permanence of commitment which is entailed in intramural support. An agency must participate actively in the continuing assessment of the significant prospects and implications of basic research, and must assure its own capability to make prompt and vigorous response to innovations made possible by the combination of discoveries in widely different fields of science. The lead time for the evolution of such discoveries into the possibility of a system development is so long that a high price can and should be paid for the earliest possible forewarning and appreciation of the significance of such discoveries. But since payoff will be infrequent, the agency must select a low level of participation in a broad range of scientific disciplines, laying its emphasis on gaining entry to the highest quality work and the most productive groups in each field of this type.

The listening-post type of activity requires a much more creative role from scientific program officers than the mere administration of a broad scientific program. They must be in close scientific communication both with their contractors and with the applied needs of their agency. Above all they must have the time to think, to put their feet up on the table and to speculate frequently on where their science and technology are going. And they must have the time to travel into the field, not only to see what is going on in the laboratories which they sponsor, but also to stimulate the thinking of their contractors and carry ideas from one to the other.

Perhaps nowhere has the value of the listening-post type of

commitment been better exemplified than in ONR's long-standing participation in the support of nuclear research. The benefits to the Navy from nuclear research have come not so much from the research results themselves as from the derivative technology, which has resulted from the fact that accelerator design and particle-detection instrumentation have continually pressed the state of the art in areas of technology which have proved important to the Navy and to the military services generally.

For example, the first high-power klystrons were designed specifically to power the Mk III linear accelerator at Stanford, increasing by a factor of 1000 the power available from such microwave generators when the development was first started in 1947. In the meantime, the linac itself has become a commercial device of considerable importance both for medicine and for industrial radiography, including incidentally the radiography of very thick objects, such as missile propellants *in situ*. The electron linac development also provided an important impetus to the more general development of microwave technology. A variety of microwave components, such as attenuators, phase shifters, and high-power windows, and new microwave measurement and calibration techniques, were developed in connection with these machines. All of this microwave technology has been of great significance in connection with modern high-power radar. Of course, this technology did not emerge from Navy-supported work alone, but the fact that the Navy was one of the earliest in the field contributed to the rapid importation of these techniques into defense technology.

Nuclear physics early generated exacting requirements in the area of fast pulse electronics, and the demand for extremely short resolving times in coincidence counting of particles has stimulated many circuit developments which have interacted fruitfully with parallel research in computer technology. The most widely used commercial oscilloscopes were developed originally to fill the needs of nuclear physicists.

A decade ago the increased use of scintillation counting led physicists to press for the development of very sensitive photomultiplier tubes. They advised the government on developmental contracts with industry to develop tubes specifically tailored to the needs of nuclear physicists. Similar tubes now find application as the most sensitive detectors of faint light in many areas of military and civilian technology.

It is doubtful whether any of these new technologies would have been developed as rapidly or as economically by any other means, because at the time the developments were started, the need for them in other areas than nuclear physics could not have been foreseen sufficiently clearly to provide the necessary focus and incentive for the development effort. This is only one of many examples of how the tools of basic research anticipate the needs of other more everyday technologies, and thus serve as a stimulus to bring forth the art and science that become available for other uses when the time is right. In my opinion the military services ought to use this technique much more frequently than they do, in order to cultivate the technologies that are likely to be needed ten to fifteen years from now.

Today nuclear physics and elementary-particle physics are challenging technology in new directions. One example is the marriage of microwave technology with cryogenics to produce higher efficiency microwave power components, an area in which an ONR-sponsored contractor is currently the leader. Another example is the development of sophisticated data processing techniques for "on-line" experimentation and "pattern recognition" of bubble chamber and other particle-detection patterns. Some of this work has been and still is ONR-supported, giving the Navy a "window" on new technologies likely to be of profound significance for future military applications, but in ways not now clearly foreseeable. As the Navy's fractional commitment to the national nuclear and high-energy physics program has decreased over the years to less than 7 percent, it must be more and more selective in its support of this area, but it seems doubtful whether its participation could drop much lower than at present and still afford it a meaningful window on the field.

The current conventional wisdom has it that nuclear and elementary-particle physics are "useless" subjects, worthy of support, if at all, only for their "cultural" value. This is why I deliberately chose the Navy's nuclear physics program as an illustration of the value of a listening-post commitment to a few of the most vital frontiers of advancing science. This is not an argument for indiscriminate support of *any* basic science on the part of mission-oriented agencies. What I have tried to emphasize is that the ecology of the scientific effort is far more complex than the naïve connections which can be made between pure science and applications before the fact. These connections are often not direct but proceed through many layers of neighboring sciences and instrumental and industrial technology. While I am all for projects like Hindsight which attempt to trace the origins of modern weapons systems, I would warn that such efforts are very likely to lose the trail just at the most interesting point when it disappears into the general scientific background and sophistication of the times.

As one reviews the history of American science and technology in the last twenty years, one cannot fail to be struck by the strategic role which ONR-sponsored work has played. In fact, when one considers its present tiny *fiscal* role in research support compared with what it was in the early days, one is surprised at its still major importance and influence. Wherever the most important advances are being made, one still seems to find ONR present with at least token support. A mere catalogue of areas in which ONR-sponsored scientists have pioneered shows how frequently ONR has been there with the right science at the right time even though few foresaw the usefulness and relevance when ONR first began to sponsor it. Let me merely list a few examples:

1. The discovery of the Van Allen belt, and the development of a research satellite which was available to take data in connection with the Starfish nuclear test when such data were needed quite unexpectedly and urgently.
2. The metallurgy of high-temperature moly alloys, which proved to be vital in connection with the Polaris program.
3. The development of the thermochemistry of titanium and its compounds, which proved to be a bible of valuable information when titanium became of practical importance.
4. The early launching of an Arctic research program, data from which suddenly proved vital when it became necessary to install the DEW line.
5. Early support of work in Bayesian statistical analysis that proved to be of great value as more sophisticated methods of detection of signals in noise became increasingly important in radar and sonar.
6. The development of the mathematical theory of diffraction and scattering of electromagnetic waves from large obstacles, which became later very important in application to the problem of minimum radar return from missiles and decoys.
7. Support of the earliest work in the field of time-shared computer systems.
8. Support of the early fundamental work on the propagation and phase stability of very low frequency electromagnetic waves, which led directly to feasibility of VLF radio-navigation systems — incidentally, a fine example of cooperation between extramural, Navy, and other government laboratories.
9. Support of fundamental work on the theory of wind-generated waves, which led eventually to operationally useful techniques for forecasting ocean waves.

10. The discovery of microplankton in the oceans, and the realization of the importance of small organisms in affecting acoustic propagation and scattering in the ocean medium.
11. The invention and development of a method for the rapid freezing of blood.
12. The support of fundamental work in oceanic geophysics, which led directly to the development of a useful geophysical navigation technique.
13. Support of the earliest work on numerical modeling of the atmosphere, which is now beginning to lead toward a practical method of numerical weather forecasting.
14. The discovery of the so-called deep sound channel as an outgrowth of fundamental investigations in underwater sound propagation by an oceanographic laboratory.
15. The development of the concept of an integrated fleet air defense system.
16. The support of early fundamental work on shock tubes and shock dynamics, which was the direct forerunner of the use of shock tubes in the study of re-entry problems and the development of practical nose cone material —a primary example of a basic research tool which, through remarkable prescience, was ready to be applied in testing and development programs when needed, even though nobody had conceived the ICBM when the work was first supported.
17. The development of the plastic cornea for eye repair, an example of assiduous and inspired follow-up on an initially fortuitous observation.

One could go on with this list indefinitely, but I think I have recited enough examples to make my point. On the other hand, we should not fall into the trap of believing that the basic idea is everything. I suspect many of these ideas would have been lost in the general noise if there had not been alert and intelligent program officers and Navy scientists who had the wisdom to appreciate their potential and see that it was further developed. The accomplishments of ONR-sponsored research must be a source of pride not just to the scientists who did the pioneer work but to the creative administrators and naval officers who cultivated their scientific gardens so fruitfully. I think too many of us in the scientific community have recently been too inclined to forget the importance of this role; and on the other side, too many in the policy-making positions of government, while giving lip service to basic science, have been too inclined to forget that scientific results cannot always be whistled up to order when needed. They have to have been brewed years before, and in military development there is nothing more costly than the basic scientific results that were not available when the time was ripe for their use.

SIX □ Future Needs for the Support of Basic Research

In the spring of 1964 the National Academy of Sciences was asked by the Subcommittee on Science, Research, and Development of the House Science and Astronautics Committee to prepare a report for Congress concerning future national needs for the support of basic research. The Academy was requested to address itself to two specific questions:

1. What level of federal support is needed to maintain for the United States a position of leadership through basic research in the advancement of science and technology and their economic, cultural, and military applications?

2. What judgment can be reached on the balance of support now being given by the federal government to various fields of scientific endeavor, and on adjustments that should be considered, either within existing levels of overall support or under conditions of increased or decreased overall support?

The President and Council of the National Academy assigned the task of preparing such a report to the Academy's recently organized Committee on Science and Public Policy (COSPUP), then chaired by Dr. George Kistiakowsky, professor of chemistry at Harvard University, and formerly Special Assistant to the President for Science and Technology in the last two years of the Eisenhower Administration.

COSPUP delegated the task to an ad hoc group whose membership was drawn about half from COSPUP itself and half from scientists outside COSPUP, including two economists. This group, instead of preparing the usual type of consensus report, chose instead to produce a series of "criticized essays" by the individual members. Each of the personal essays was reviewed in face-to-face sessions with all the other members of the group, but the author was free to accept or reject the criticisms and suggestions as he saw fit.

The following chapter is a very slightly edited version of my contribution to that report, which appeared under the title "Basic Research and National Goals."

In common with most of the other essays in that volume, my paper gives no specific answers to the questions posed by the House committee, but rather attempts to set forth some of the considerations and criteria that should be used in arriving at tentative answers on a continuing basis.

INTRODUCTION

The two questions posed by the House committee are exceedingly difficult to answer in any precise quantitative way. The general approach taken by this paper is that the answers can only be arrived at by successive approximations. We thus try to suggest some of the considerations and some of the mechanisms of choice that ought to be considered in determining levels of support for science.

I begin my paper discussing some of the problems involved in interpreting research and development statistics. Since current statistics must provide the basis for any future planning for science it is important that the limitations of these statistics be fully understood.

The second section deals with some of the reasons why the support of basic research is considered to be in the national interest, and why this support must be primarily a federal responsibility. In this section I suggest some possible guidelines for future overall support of academic research.

In section III a conceptual scheme for considering the "science budget" is suggested. This involves an attempt to separate the requirements of big science from those of the individual investigator in the university. It suggests that the problem of relative allocation to fields is not one to be centrally determined, but rather a question of setting up suitable mechanisms for continuing decentralized choice. This section is concerned mainly with academic research.

The fourth section attempts to describe the difference between academic research and organized institutional research, and to explain the different mechanisms of choice and criteria that should apply to the latter as compared with the former.

I. SOME REMARKS ON RESEARCH AND DEVELOPMENT STATISTICS

Since much current discussion of federal spending on science is based on financial and manpower statistics, it is important that the meaning and limitations of these statistics be fully understood. A recent report of the Organization for Economic Cooperation and Development has remarked that most countries have better statistics on poultry production than they do on the activities of their scientists and engineers. To some extent this is inevitable since the product of scientific activity is an elusive entity that defies measurement. Especially in basic research we have nothing but historical analogy to go on in evaluating the worth of the product, and even in purely scientific terms the value of any given piece of work often does not become fully apparent until several years after it is published. In many cases an unsuccessful experiment may have more lasting value than a successful one. A classic example is the famous Michelson-Morley experiment, which failed to detect the absolute motion of the earth through space and led directly to Einstein's formulation of the theory of special relativity, but not until many years later.

As a result of these features research activity is not very amenable to the ordinary methods of economic analysis. We can measure the "inputs" in financial terms or in terms of "professional man-years" of effort, but we have no comparable currency in which to measure the "output." We can see the continuing growth of our economy as primarily a product of technical innovation, but until very recently little of this innovation was clearly connected with organized research and development. No striking acceleration of economic growth has accompanied the dramatic growth of organized research in recent years. This is not very surprising in view of the large average time lag between research discoveries and their application. On the other hand, the sectors of the economy showing the largest percentage growth rates are in many cases those most heavily dependent on modern research. All the advanced industrial countries devote about the same proportion of their national income to civilian research and development. Thus we have no "controls" by which we may judge what would have happened to economic growth if there had been no research and development, nor do we have a way of measuring the relative importance, economically speaking, of the research relative to the development. Indeed, there is no economic payoff from research until it is incorporated in some kind of product, service, or process, and this won't happen to current research results, for the most part, for many years. Thus research, and particularly basic research, is a speculative investment in the relatively long-term future; its economic payoff has a longer incubation time than any other form of investment, except possibly education.

On the other hand, there are certain things that can be said about the current economic benefits of technical proficiency to the United States. For one thing, this country has an exceedingly favorable balance of trade in "technical know-how," as measured by international payments for royalties, licensing agreements, and management fees. Such payments net nearly half a billion dollars a year, and payments to the United States exceed, by a factor of nearly 5, payments from the United States to all other countries. As another example, analysis of our exports clearly reveals that the proportion of products from industries that may be classed as "research intensive" is very much higher in our export trade than it is in the gross national product as a whole, suggesting that it is the industries based on technical know-how that generally compete most effectively in world markets. Analysis of the exports of other advanced nations indicates a similar bias toward products and services based on research. On the other hand, one must interpret these figures with some caution, since they must relate primarily to technical advances that took place before the present high federal investment in research and development, and since technical progress in Europe and Japan was heavily retarded by the effects of World War II and its aftermath.

Because of the absence of valid economic measures for the product or benefits of research and development we are forced to measure it essentially in terms of its economic inputs, with the implicit assumption that in some sense the output will be proportional to the input. In terms of inputs, one thing is clear: research and development probably constitutes one of our fastest-expanding forms of economic activity. Nevertheless, one must regard statistics of the past with great caution. Even within a span of a few years, there has been a tendency to include more and more activities under the category of research and development that were formerly looked upon as part of production or design or, in the military field, procurement. A few years ago, as a result of a reorganization, the category of research and development in the Defense Department was changed to "research, development, test, and evaluation." This placed the dividing line between development and procurement much further along in the weapons system cycle than had formerly been the case. Now sample production runs of weapons for evaluation, and the costs of expending them under simulated service conditions, are treated as part of research, development, test, and evaluation. Apart from this effect, which caused a discontinuous 20 to 30 percent jump in the apparent research and development budget of the Department of Defense, the general popularity of research and development probably resulted in a good deal of redefinition of many technical activities. Thus the growth of research and development in the last decade, while substantial, is probably not as rapid as indicated by the raw statistics.

On the other hand, there is much activity of a highly technical nature in the federal government which, while not classified as research and development, requires the participation and supervision of people with advanced technical training and experience. Many of the services performed by government involve the collection of technical data on a more or less routine basis. Examples occur in weather forecasting, hydrographic and geological mapping, and collection of economic and population statistics. That the function of the federal government in our society is highly technical is indicated by the fact that nearly 50 percent of the professional civilian employees of the federal government are scientists, engineers, or health professionals, and the three highest grades of the civil service are even more heavily populated with people with technical backgrounds.

Similar problems arise when one talks about specific classes of activity, such as basic and applied research. In the first place, the motivations of the man who does research can, quite legitimately, be different from the man who supports it. In the second place, some basic research involves the design, construction, and operation of very large and complex equipment. The motivation for acquiring this equipment may be purely scientific, but much of the activity accompanying its design and use is indistinguishable from the more applied kinds of engineering or production. Thus, for example, in fiscal year 1964 the National Science Foundation reported a federal investment of about $1.6 billion in basic research. It turns out that nearly half of this amount was spent by the National Aeronautics and Space Administration and that approximately 80 percent of the National Aeronautics and Space Administration expenditure was for the design and procurement of scientific space vehicles, the operation of tracking ranges, and payments to military missile ranges for putting the vehicles into orbit. A significant part of the oceanography budget goes into simply keeping research vessels at sea, without any science. The operation of a large particle accelerator requires annually something like 10 percent of its capital cost, or perhaps as much as 30 percent if one includes the cost of continued updating of the equipment. Similar figures can be quoted for large optical telescopes or arrays and "dishes" for radio astronomy. These are operating costs that are required simply to make a facility available, with no consideration of the additional costs of the actual science to be done.

Why is it necessary to stress these logistic costs of research? Since they are incurred for the purpose of achieving basic research results, they are legitimately chargeable to basic research. Nevertheless, the impression conveyed by statistics that include such supporting costs can be quite misleading. A basic research budget that rises annually by 15 percent may appear to be adequate or even generous, but if most of this cost increase is merely to ensure the availability of certain new facilities, then the increased budget could actually be supporting the activities of fewer scientists. The situation would be a little like building a new department store that was so expensive to keep open that it was necessary to fire all the salesmen. This is not an academic issue. Much of the planning for new research facilities that took place in fiscal years 1962 and 1963 was based on an implicit assumption of continuing expansion of research budgets. Now, in fiscal years 1964 and 1965, when these facilities are just coming into operation, the expenses of merely making them available — without any science — are confronting fixed or even declining operating budgets for basic research. The political embarrassment that would attend not using a facility already built makes it inevitable that the facilities are made available anyway, usually at the expense of the individual scientist who does not have large fixed costs. A recent calculation indicates that if the budget for oceanography continues to stay level, the cost of operating ships already planned but not yet completed will eventually consume almost the entire research budget. A similar situation appears to be developing in low-energy nuclear physics, and with respect to university computing facilities. In nuclear physics, for example, expenditures for facilities doubled between fiscal years 1962 and 1964, while operating expenditures increased only slightly and actually decreased in the university sector between 1963 and 1964. The point I am making is that simply to look at total budgets for basic research, or even their annual increments, can be highly misleading unless one knows something about the fixed availability costs that have been built into the program by past commitments for capital facilities. Where large availability costs are involved, the relation between research output and dollar input can be highly nonlinear, and hence measurements of basic research activity by dollar inputs can give a misleadingly reassuring impression as to the adequacy of support. Unfortunately, our present methods for collecting and classifying statistics on research expenditures are not sufficiently refined to reveal problems of this sort, or to draw clearcut conclusions about the current situation. Subjective opinions of many individual scientists and research groups indicate that support for the individual investigator is becoming increasingly inadequate relative to his needs, but it is hard to prove this quantitatively, and even harder to establish that it is due to past commitments for facilities.

Classification of research into basic and applied can also be misleading as to the type of manpower required. In the space example, a single experiment may involve the services of hundreds of technicians and skilled workmen, whereas only four or five scientists may be involved in the actual design of the instrumentation package and the analysis and interpretation of the data. The same amount of money in another field of research might finance the activities of fifty highly trained scientists. This issue is an important one because it is sometimes claimed that there is more money for basic research than the really competent people available to do it can spend wisely. This could be true, but I submit that it is a judgment that cannot be made in terms of total available funds, but only on a project-by-project basis. Two or three competent scientists can in some areas of research wisely command or direct the activities of a large number of less highly skilled people. In fact, one effect of increased research funds is that many scientists are able to buy from industry equipment that they would otherwise have to design and build themselves. The dollar input in their research is much larger than it would otherwise have been, but this does not necessarily mean that the research is more "expensive" if measured in terms of the research results obtained. The capital investment may not only enable the scientist to obtain more results for the same effort, but also may permit him to choose a much more significant problem or to obtain a much more conclusive answer. Just as capital investment embodying new technology improves the productivity of ordinary labor, so does it improve the productivity of scientific effort. Unfortunately, since it is the only thing that is quantifiable, there is a tendency to measure research in terms of man-years of effort or in terms of output of publishable papers. While the latter is certainly much more significant than the former, there is still too wide a variation in the information content and quality of scientific papers for paper publication to provide an accurate measure of research output.

Another statistic that is often quoted has to do with academic research. For a long time it was common practice to report only the total research and development support going into universities. However, in the postwar period many universities undertook the management of large applied laboratories or basic research institutes. Some of these, like Los Alamos, were remote from the campus and had no visible intellectual connection with the parent university. However, such clear-cut cases were the exception; usually the relationship to to the university was closer, as in the case of the Radiation Laboratory at Berkeley, the Cambridge Electron Accelerator, or even the Lincoln Laboratory of MIT. It has now become customary, however, to classify such organizations as federally financed research centers and exclude them in reporting the support of research in "universities proper." Nevertheless, there are many such organizations that employ faculty members part time and participate in the training of graduate students. Other organizations, such as the Brookhaven National Laboratory, the National Radio Astronomy Observatory, or the Kitt Peak National Astronomy Observatory, are not classified with universities at all, but nevertheless provide important facilities for university "user groups," including significant numbers of graduate students and faculty on temporary assignment. Conversely, there are some research activities within "universities proper" that are little more than research institutes with rather minimal intellectual connection with the rest of the university. The point here is that the line between "academic" and "nonacademic" research in universities — between universities proper and research centers — is not a sharp one if measured by involvement in the educational process. Yet, with respect to federal research and development investment, the research centers account for something like 40 percent of all university research activity. With current emphasis on the connection between basic research and graduate eduation, there is a danger in completely eliminating the research center statistics from the overall picture, with the implication that the elimination or downgrading of such activities would have no effect on the educational function of the universities. In some cases this might well be so, but in others it would not be. There is equal hazard in the converse assumption that all the research funds going to universities proper are in support of graduate education and therefore required to maintain the quality of graduate training; unfortunately, we have discovered no quantitative way to measure the educational relevance of research funds.

Another statistic that may be misleading is the separation of federal funds into contributions to "research" and "education." Thus, on the one hand, in reporting federal research and development funds in universities proper, fellowship funds, research training grants, and certain types of institutional support are usually omitted, despite the fact that many of the individuals who receive stipends under such programs are actually engaged at least part time in research or in the supervision of student research. It is clear that a significant proportion of such funds contributes to the progress of research in universities. In the National Institutes of Health they amount to about 30 percent of all the funds contributed to universities, although they are less significant for other agencies. On the other hand, a very large proportion of the funds designated as "research" actually provide stipends for graduate students and postdoctoral research associates who, while engaged in research, are also receiving training. Indeed, since research experience is believed to be the most important and valuable part of advanced training in science, the separation between research and education funds is bound to be rather arbitrary and artificial.

Even the classification of research funds into federal and nonfederal may be highly misleading. For example, procurement contracts in defense, space, and atomic energy permit business organizations to charge a small fraction of their independent research activity to procurement overhead and also allow technical work in connection with the preparation of development proposals, including unsuccessful proposals, as an overhead item. It has been estimated that the total funds channeled to industry in this way amount to close to $1 billion, about the same amount of money as flows from the federal government into universities proper for research, basic and applied. Yet this money is classified in the statistics as being financed by the private, not the public, sector. A good deal of private research is also financed out of the profits of military and space procurement. The proper classification of these activities is hard to decide. In the sense that the basic resource-allocation decisions are made in the private sector, regardless of the source of funds, the activity is correctly classified as private. On the other hand, the government does exercise some surveillance over the expenditure of part of these funds. Furthermore, the extent and scope of the activity are strongly conditioned by decisions in the public sector.

In considering research in the university sector it is often forgotten that, in practice, the salaries of faculty members engaged in research are paid largely by the university out of its own sources of funds, and are not a charge against federal research and development budgets. This is in contrast to federally financed research centers and to research in private industry, where the federal government is routinely expected to bear the full costs. In addition, the universities make a major contribution in the form of unreimbursed indirect costs, estimated to exceed $60 million annually. In a sense that does not apply to any other sector to the same degree, the federal contribution to university research is a contribution to a shared activity rather than procurement of a service at cost. Any increase in the federal contribution to university research thus generally reflects an increased contribution from other sources as well.

In considering the totality of federal research and development activities, there appears to be no unique way of breaking down expenditures into their significant components. Except possibly in the area of specific hardware development, most federal research expenditures serve several purposes simultaneously, and most scientific activities relate to more than one traditional disciplinary categorization. The network of communications and organization in the technical community is so dynamic and complex that it is difficult to capture in a statistical snapshot at any one point in time, and even harder to characterize by fixed statistical categories over a period of time. In my personal view the most reliable and useful statistical categories are those that relate to institutional arrangements, such as universities, federal research centers, and scientific departments or schools, rather than to such categories as basic and applied or to the various traditional scientific disciplines.

II. WHY SHOULD THE FEDERAL GOVERNMENT SUPPORT BASIC RESEARCH?

The House committee has asked at what level basic research should be supported in order to maintain our present position of leadership. As background for answering this question it is necessary to inquire why the federal government should support basic research in the first place, and what functions basic research serves in our society.

One can recognize four distinct functions of basic research, some of which also pertain to certain types of applied research. They are: cultural, economic, social, and educational.

Cultural

Basic scientific research is recognized as one of the characteristic expressions of the highest aspirations of modern man. It bears much the same relation to contemporary civilization that the great artistic and philosophical creations of the Greeks did to theirs, or the great cathedrals did to medieval Europe. In a certain sense it not only serves the purposes of our society but *is* one of the purposes of our society. Science and technology together constitute the distinctive aspect of American culture that is most admired and imitated in the rest of the world, and I believe this admiration is connected with more than the economic and military power that derive from technology.

The attitude of the general public toward the space program suggests that this cultural aspect does enjoy a degree of public acceptance. While it is true that much of the public supports the space effort because it feels in a somewhat vague way that it is connected with military power, nevertheless there is a genuine sense of identification with the adventure of exploration into the unknown. To the scientists it may seem naïve that the public should identify the space program, especially the man-in-space program, with science. To many, but by no means all, scientists the relative emphasis on the lunar-landing program appears as a distortion of scientific priorities and of intellectual values. Is manned exploration of the near solar system really worth a thousand times as much as probing the secrets of distant galaxies or the dramatic and intriguing quasi-stellar energy sources? Nevertheless, public acceptance of the space program must be regarded as in some sense a vote of confidence in intellectual exploration as such and a recognition of the desirability of public support for such exploration. This recognition is, by itself, a new political phenomenon, and may represent only the first step toward a wider and more informed public recognition of the desirability of social support of intellectual exploration for its own sake.

Any statement of a cultural motivation for the support of basic research raises, of course, much more serious issues of political philosophy than the other motivations listed. Why basic science but not art, music, and literature? Why not research in the humanities? If we support science for cultural reasons, how can we tell how much is enough? I think the only definite answer that can be given to these questions lies in the nature of science as a system of acquiring and validating knowledge. Science — especially natural science — has a public character that is still lacking in other forms of knowledge. The results of scientific research have to stand the scrutiny of a large and critical scientific community, and after a time those that stand the test tend to be accepted by all literate mankind. Outside the scientific community itself this acceptance tends to be validated by the practical results of science. If it works it must be true. There is no question that the successful achievement of an atom bomb provided a certain intellectual validation for nuclear physics, quite apart from its practical value. Part of the public character of science results from the fact that it is always in principle subject to independent validation or verification. It is like paper money that can always be exchanged for gold or silver on demand. Just because everybody believes that he can get gold for paper, nobody tries; so the public seldom questions the findings of science, just because it believes that they can always be questioned and revalidated on demand. This is much less true of other forms of knowledge and culture, which may be of equal social importance but are more subjective and more dependent on the vagaries of private tastes and value systems. It is just because science is a cultural activity generally believed to transcend private value systems that it becomes eligible for government support where other forms of cultural activities are not. The system of indirect public support through tax exemption has been used in the United States successfully to support cultural activities in areas where there is no consensus of values or tastes. This is possible because, although public funds are used, actual decisions as to what will be supported are left in private hands. It may well be that this situation should be regarded as temporary. Direct government support of other forms of cultural expression is generally accepted in advanced countries other than the United States.

The basic difficulty with the cultural motivation for federal support of basic research is that it does not provide any basis for quantifying the amount of support required. The amount of basic research that should be supported for purely cultural reasons is certainly a fraction of what should be supported for other reasons. It is currently believed that the talent for really creative basic work in science is exceedingly rare. I believe there is a most creative minority, possibly not more than 5 percent of all the active basic research scientists, who should receive support for their work for no other reasons than their demonstrated capacity for original and creative work. This highly selected group of people might be provided with some minimum level of research support with no strings attached. They would simply be backed up to some level, say $20,000 to $30,000 per year, to work on anything they thought worth doing. If they needed more than this, then their requirements would have to be justified in competition with others in terms of their specific proposed work and for other than purely cultural reasons. I believe the government could reasonably commit something of the order of $100 million[1] a year to this type of completely freewheeling research expenditure.

It must be remembered, however, that the work of this most creative group cannot be regarded as independent of the more run-of-the-mill kinds of research, as is sometimes implied. Important discoveries have sometimes been made by individuals who never did anything else of significance in their careers. The brilliant generalizations of giants often rest on the painstaking accumulation of data by less gifted individuals. The relative importance of brilliant and intuitive insight as compared with the more pedestrian hard work will vary from time to time with the circumstances of particular fields. For example, the progress of mathematics and theoretical physics is probably much more dependent on the insights of a few leaders of extraordinary ability than is the progress of experimental physics or chemistry. One cannot support only the geniuses and expect that science will continue to progress as though the workers in the vineyard were superfluous. However, it is certainly true that more than a merely cultural motivation should be required to justify the support of other than the few most highly gifted.

For the sake of its position of leadership it is essential that

[1] Some scientists may derive their support through working in close association with the outstanding 5 percent; thus it is not legitimate to extrapolate the $100 million for the 5 percent to $2 billion for the total pool.

the nation be prepared to invest heavily in equipment and facilities which place a few of its most talented groups at the "cutting edges" of modern scientific advance. No matter how talented the people, facilities that are second best are likely to leave them in the position of verifying exciting discoveries made by somebody else. The pre-eminence of the United States in nuclear physics owes much to the brilliance of its workers in this field, including many imported from other countries, but it owes even more to the superior equipment that generous federal support, good planning, and high-class engineering have made possible. United States pre-eminence in many fields of science reflects not only the intellectual vigor of its scientists but also the excellence of its industrial base.

The United States has led the world in discoveries in optical astronomy almost since the turn of the century, and this is largely attributable to the foresight of some of the great private foundations that supported American astronomers in the construction of better instruments than existed anywhere else in the world. By constrast, in radio astronomy, despite a large investment, American instruments are inferior to some in Britain and Australia, with the result that the United States does not enjoy the clear lead in this field that it does in optical astronomy, despite the fact that the detection of radio waves from space was originally an American discovery.

Supporting basic science for purely cultural reasons, of course, pays dividends in other areas such as national prestige and the intellectual respect of the most influential groups in the rest of the world. Thus the purely cultural motivation supports the power and influence of the United States in the world and adds to the self-confidence of its own people. Nevertheless, paradoxically, supporting science solely for reasons of national prestige usually tends to corrupt it by distorting its scientific objectives and priorities, and thus ultimately to defeat the prestige objectives as well. This is generally an area where virtue is its own reward.

Economic

There is now general acceptance among economists of the importance of technological innovation for economic growth. To an increasing extent such innovation depends upon the results of basic science, although the degree to which this is true is difficult to quantify. To an increasing degree also there is a disposition to regard organized research and development as an investment in new knowledge equivalent in some sense to the investment in fixed capital. Indeed, most capital investments incorporate some measure of technological innovation. According to some economists the rise in capital-to-labor ratio accounts for only a small part of increases in productivity; about 50 percent is ascribed to other factors lumped under the general heading of "technical progress," which probably incorporates about equal parts of research and education as well as such factors as managerial and marketing innovations. There is also general agreement that in a market economy the allocation of resources to the advance and spread of knowledge will tend to be less than the optimum required for maximum efficient long-term growth of the economic system as a whole. Moreover, the further removed research is from ultimate practical application the less likely it is to be supported in a market economy without either direct public subvention or private support induced by special tax incentives, which is also a form of public support. Thus, there appear to be strong economic reasons for federal support of research, and especially basic research.

In comparing the United States with other advanced industrial countries one finds that, if one sets aside military research and development expenditures, our investment in research is about the same in terms of percentage of national income as that of other nations, including Japan, the Netherlands, the United Kingdom, Sweden, West Germany, and France. It is noteworthy that federal support of basic research in universities is a smaller fraction of total university basic research than in any other advanced country. This is, of course, because the United States has no federal university system, and also because it relies much more heavily than other nations on indirect public support via tax deductions for private contributions. The fact that the federal share of total research support in universities has increased is attributable largely to the increase in applied research, chiefly in the medical and engineering areas. Thus, in relation to our national investment in higher education, it does not appear that the federal contribution to basic research in universities is in any way exceptionally large or increasing at a disproportionate rate.

Since World War II there has been increasing recognition of the potential economic benefits of supporting science on its own terms without any commitment to specific applications. Politically, however, this commitment has always been made with some reserve. The National Science Foundation, the only agency with a clear mandate to support basic research as such, had a long struggle to come into existence, and an even longer struggle to attain a significant budget for research. Even today it accounts for only a little more than 10 percent of the support of research in universities proper — nearly 20 percent of the truly basic research. It also accounts for about 10 percent of all federally supported basic research. On the other hand, Congress has been quite liberal in permitting the mission-oriented agencies to support basic research related to their missions, and the interpretation of mission-relatedness has been reasonably broad. If it had not been for this fact, U.S. science would not have attained the reputation for world leadership that it enjoys today.

In several fields federal support for mission-related basic research has been of decisive importance for U.S. technological leadership, even in the field of civilian applications. Although the transistor was invented in private enterprise, federal support for university solid-state research played an important role in creating an environment in which the transistor could be rapidly exploited and developed. Federally supported research also greatly accelerated the development of high-speed computers, and much of the pioneering work on computers was done in universities. Federal support of aeronautical research, largely in inhouse laboratories, was important for U.S. leadership in the development of modern civilian aircraft. Undoubtedly, federal support for basic research in the medical sciences and biochemistry has accelerated the development of new drugs by industry. Support by the Atomic Energy Commission of basic nuclear research that was not obviously relevant to weapons or nuclear power has been largely responsible for the maintenance of U.S. leadership in this field.

On the other hand, in only three fields — agriculture, mineral resources, and civilian nuclear power — has the federal government explicitly supported applied research aimed at development of the civilian economy.

Social

In many areas, including public health and national defense, there is a recognized federal responsibility. In these areas the federal government has generally been quick to utilize research in support of its missions, including a substantial amount of basic research. In fact, for the most part, basic research support has tended to derive from these special missions rather than from any overt policy concerning the desirability of social support for research. More recently, beginning with the National Advisory Committee for Aeronautics in 1920 and extending through the Atomic Energy Commission and the National Aeronautics and Space Administration, the government has recognized a special responsibility for exploiting certain advanced technologies in the national interest. In these cases it was recognized that the technologies were sufficiently new and unappreciated so that they would not be adopted and adequately supported as part of the missions of existing federal agencies or private institutions. They needed hothouse cultivation, as it were, before they could grow and mature on their own. In each example of such an agency, however, there was a strong military overtone to the justification; it is doubtful whether the National Advisory Committee for Aeronautics, the Atomic Energy Commission, or the reincarnation of the National Advisory Committee for Aeronautics in the National Aeronautics and Space Administration would ever have been justified without a quasi-military incentive. However, once there, their additional roles in economic growth gradually came to be appreciated.

It is clear that with increased urbanization and industrialization, our country is developing a number of problems that can only be faced on a national basis — for example, education, air pollution, water resources, weather forecasting and control, pesticides, radioactive wastes, public recreation, natural resources, air traffic control, highway safety, and urban transportation. The degree of federal responsibility in these areas will always tend to be a matter for political debate, but there is greater consensus that the federal government has a responsibility for seeing that the foundations of knowledge are laid in these areas than there is that it has an operational responsibility. Research related to these social goals tends to be recognized as a federal responsibility even when operation or regulation is delegated to the state or local level or to private enterprise. If applied research for these purposes is a federal responsibility, it is clear that the basic research that underlies it must also be recognized as a federal responsibility. Except in the areas of health and national security, however, there is still little appreciation of the contribution that uncommitted basic research can and should make to these goals. What is called basic research in many areas of federal civil responsibility is still rather narrowly oriented in terms of obvious relevance to the immediate goal. Such oriented basic research is vital, but not sufficient. The rather rigid interpretation of relevance to mission that exists in the research in the older civilian agencies is in sharp contrast to the broader interpretation that is followed in national defense and health.

The difficulty with this motivation for federal basic research is that criteria for the amount and character of basic research that should be supported in connection with social goals is difficult to establish. Clearly it is proper that research as a whole in these areas should compete on an equal basis with alternative means of achieving the same goals. Perhaps the only reasonable criterion is to relate the basic research effort of an agency to its total applied or development effort, possibly in terms of some percentage of the applied effort. Any such criterion, however, should involve some smoothing of fluctuations to take into account the longer time frame of basic research. The fractional effort on basic research will inevitably be strongly dependent in the breadth of the mission of an agency and on the magnitude of the total effort and its degree of dependence on relatively new or recently discovered scientific knowledge. I would suggest that in many instances 10 to 15 percent of the applied effort might be a good rule of thumb for the basic research effort. However, it is difficult to mount a viable basic research effort when the applied research is too fragmented into small units, as it is, for example, in the Department of the Interior, the Department of Agriculture, and the Department of Commerce (except for the National Bureau of Standards). In such cases it might make more sense for these departments to "task" the National Science Foundation with basic research in certain broad areas of relevance to the total mission of the department. It also seems rather important that not all the research, either basic or applied, be inhouse. Exclusively inhouse research often appears to be more efficient in the short run, since the people involved can be more closely channeled into research areas that meet the short-range requirements of the mission, but in the long run a purely in-house research effort tends to cut the agency off from the scientific community. Not only is the scientific and educational community unaware of its problems, but its own people lose awareness of the opportunities that new developments in basic science present in the applied research it is doing. It always tends to define its own subject matter too narrowly.

Education

The intimate connection between basic research and graduate education has been repeatedly stressed in recent years. In engineering, medicine, agriculture, and several other areas, applied research is equally important as advanced training, and there is danger that this fact may be forgotten in identifying the universities too exclusively with basic research. In particular, there is a tendency in the universities to regard the application of science as a lower order of intellectual activity than pure science, an attitude that tends to impede the healthy flow of talent between basic and applied science, which has been one of the characteristic features of American science contributing to its vitality. On the other hand, it is true that even in applied research the universities ought to focus on the longer-range goals, the things that are likely to become economically viable several years away, and that have the greatest generality in application. Research apprenticeship is the most essential part of graduate education beyond the master's level, whether it be in pure or applied science.

There is a broader sense in which research activity contributes to education. Research itself is defined as "learning work" — the production of new knowledge. While much of this knowledge is made explicit and public by publication in the technical literature, the individuals engaged in advancing knowledge acquire skills and perspectives that greatly transcend the sum of the information appearing in their publications. The contribution of a Fermi or a Von Neumann to our society is far greater than represented in the bound volumes of their collected works or even in their influence on their students. A great scientist becomes a teacher of his whole culture. The people who devote most of their lives to research become a national human resource, available in emergencies to turn their attention to many problems outside their own immediate fields of interest. The rapid application of microwave radar during the early years of World War II was largely the work of nuclear physicists, even though the basic invention had been conceived several years earlier by engineers in government laboratories. What was needed for the exploitation, however, was not just the invention itself but a whole complex of experience with advanced electronic techniques and with the integration of these techniques into an operable system. The nuclear physicists who had been working with accelerators possessed this kind of experience, and were able in an emergency to turn it to military applications. Through the decade of the 1930s, they had been unknowingly educating themselves, in a sense, for just this moment. It is doubtful whether any explicit or conscious form of education would have been as effective as their own continuing involvement in basic science. What applied to radar was even more evident in the case of nuclear weapons, since only those previously engaged in nuclear research, chemical kinetics, radiochemistry, and other fundamental fields had the accumulated skills necessary to proceed with projects in this field. The contribution of the engineering management skills of American industry — especially of the chemical engineering industry — was also indispensable, but without the intellectual leadership and vision of the basic scientists the project would neither have been undertaken nor carried to a successful conclusion. The development of the electronic computer in the early postwar years owed much to the high-speed electronic-circuit techniques in which nuclear physicists had trained themselves in order to sharpen the tools of their own basic research.

Not all individuals who receive advanced training in basic research remain in basic sciences. Some enter basic research in industry or government but then move on to applied science or technology in the course of their careers, often following a basic-research development or technique through into its applications. Many techniques now common in industry, such as high-vacuum technology, low temperatures, X-ray diffraction, spectroscopy, nuclear-reactor physics and neutron instrumentation, radioisotopes, electron microscopy, had their origin as techniques of basic research. Hence, there is a demand in industry for people trained as basic scientists in such fields who then find their careers in applications. The staffing of major new technological or scientific programs such as nuclear power and nuclear weapons, space research, oceanography, or atmospheric sciences has come from people with original training in basic research in physics, chemistry, mathematics, or biology. This transfer of people forms one of the major vehicles for the translation of basic science into applied science and technology, as well as for the creation of new hybrid disciplines. Thus, basic science tends to be a net exporter of people into other more applied fields of science or into technology. Too little is known, actually, about the transfer of people between fields and the influence of people receiving basic research training in one field on the development and success of other fields. It seems clear, however, that the training of people in the most advanced techniques and concepts of basic science not only is beneficial to the development of basic research itself and of graduate education but also has an important influence on the development of technology and of new industry.

Other individuals trained in basic science may choose basic research as a career but make important contributions on a part-time basis to technology and applied science. Von Neumann, a pure mathematician, formulated one of the key concepts of computer organization. Fermi, a pure physicist, conceived the idea of the nuclear chain reaction and played a leading role in its practical exploitation.

Many key ideas of military technology in the 1950s benefited from important contributions from basic scientists acting as amateur weaponeers. These people brought fresh viewpoints, new combinations of skills and techniques, and a broad vision of the potentialities of science to the weapons business. This contribution was often traceable to their basic research background. These contributions are an incidental benefit deriving from the vigorous support of basic research by the federal government, but they have played a significant role in the maintenance of United States pre-eminence in military technology.

Between the graduate student working as a research apprentice and the professor or laboratory scientist working at the frontiers of knowledge there has grown up a new group, the postdoctoral research staff, who also participate in the educational process, both as students and teachers. Such people have no formal part in the educational process; nominally, they are just research workers. They do not earn degrees, and they do not teach classes. But they both help in the detailed guidance of graduate students and deepen their own knowledge in their chosen fields. Many university departments now have as many postdoctoral fellows as graduate students, largely supported out of federal research grants and contracts. Most of them stay only a few years and then move on to more permanent academic posts as full-fledged teachers. Because of their lack of formal academic status, we know very little about this group, although their support constitutes a very significant fraction of the total research money going into universities. In some other countries, notably Sweden, the United Kingdom, Japan, and the U.S.S.R., there exists more formal recognition of the status of the postdoctoral student in the form of the D.Sc. degree, a sort of superdegree awarded on the basis of a body of significant contributions to the scientific literature.

The advantage of discussing the educational purposes of basic research is that this is the criterion for research support that is easiest to quantify. To an increasing degree U.S. policy has been evolving toward a consensus that, at least in science, society as a whole should be prepared to underwrite the opportunity for every individual to carry his education as far as he is willing and able to go.

Thus, by extrapolating long-term cultural trends, we are able to estimate fairly well how many people will be seeking graduate education in science and engineering during the next decade. The people who will do so are already in high school and college today, so there is not too much guesswork involved. The estimates of annual growth in the number of graduate students vary between 5 and 10 percent. The number has been about 8 percent for the last two years, but for the most part these students have not yet entered the research phase of their graduate study, so the full load on university research budgets has not yet been felt.

One can use the above figures to set a floor to the university research support required in the next ten years if one makes certain plausible assumptions, as follows:

1. The percentage of college graduates seeking graduate education in science will remain relatively constant or grow slightly.
2. The student-professor ratio will remain about the same as at present.
3. The ratio of postdoctoral students to graduate students will not grow beyond its present value.
4. The percentage of the total budget going into the support of large facilities, either construction or operation, or

the support of research institutes relatively divorced from teaching but still in "universities proper" will remain about as at present.

5. Because of the increased sophistication of research—including such items as more automatic data taking and data processing, greater use of computers, and greater availability of sophisticated instrumentation for purchase rather than local construction—the cost of research per man-year of research effort will increase at an annual rate of 5 percent in constant dollars.
6. The contribution to research in universities from state, local, and private sources will increase at the same rate as the federal contribution, so that the federal share will not change.

With these assumptions one arrives at a university research requirement that rises at the rate of 13 to 15 percent annually. It is interesting to note that this figure agrees rather closely with projections of requirements for the optimum scientific development of selected fields of science made by several committees of the National Academy of Sciences, set up under the Committee on Science and Public Policy.

It is important to note that almost all the assumptions in the above projection are conservative. For example, during the past ten years, with relatively little growth in the number of graduate students, the research investment per Ph.D. granted one year later has increased by a factor of 2.5. This represents an increase of 10 percent a year on a per-man-year cost basis, nearly twice what is assumed for the next decade. We are not sure of all the reasons for this growth. We suspect it is due mainly to a change in the character of universities that has been going on for the last thirty or forty years, and that was probably accelerated by the availability of federal research funds. Research has become an increasingly important part of the purpose of more and more American universities, as it has been of European and British universities for many years. Although university faculties have probably increased by less than 30 percent during this period, Ph.D. faculty has more than doubled. Furthermore, the population of postdoctoral research associates and to some extent the growth of research institutes with permanent research staff or research professors have caused research costs on a per-Ph.D. basis to rise rather rapidly, especially in the biomedical field. However, it is to be noted that, because of the upward trend of salaries in the last ten years, the normal annual increase in cost per man-year of scientific effort has been more like 7 than 5 percent. The difference between 10 and 7 percent, or 3 percent, thus represents the cost of the general change in the character of the research economy of universities and is not really dramatic. In the above projection we are assuming essentially that this long-range cultural trend will stabilize, a somewhat doubtful assumption. On the other hand, it is also true that the last decade was a period of rapid inflation of academic salaries, which had fallen seriously behind the cost of living during the postwar inflation. Academic salaries, at least for scientists, have now reached approximately their prewar position, and it is doubtful whether the inflation of the past decade will continue. Easing off of defense development expenditures may also take some of the inflationary supply-demand pressure off scientific salaries generally, especially in view of the projected increase in the supply of Ph.D.'s. The assumptions regarding the postdoctoral population are also probably conservative. On the other hand, this is the part of the academic research budget with the greatest flexibility; its size tends to be adjustable to the total funds available. A disproportionately large fraction of postdoctoral staff is probably of foreign origin, although many of them ultimately remain in the United States and take academic or industrial posts. With respect to the increased research orientation of university and college faculties, the assumptions are almost certainly conservative. As older professors oriented primarily to classroom teaching retire, they are likely to be replaced by younger men who expect to combine teaching and research. To an increasing degree it is expected that undergraduates will participate in research. Many formerly purely undergraduate institutions are talking about expanding into graduate work, if only to attract faculty of the requisite competence to maintain the quality of their undergraduate programs. Several areas of the country, especially the South, are just at the beginning of recognizing the importance of research in the functions of a university. These expectations are not really taken into account in the estimate of 15 percent a year given above. They will not be satisfied unless one of several things happens:

1. Research funds for universities are increased faster than 15 percent a year.
2. There is a substantial cutback in support of new major research facilities at universities and support going to post-doctoral associates and career research staff.
3. Other sources of financial support for research become available, possibly as a result of tax incentives to induce greater contributions to university research from industry, or special federal programs to encourage matching research funds from states.
4. The declining posteducation job market for scientists and engineers induces college graduates to seek other careers outside the technical fields, so that present estimates of the demand for graduate education are grossly inflated.

In my opinion 3 and 4 appear highly unlikely. It is remarkable, in fact, that the nonfederal contribution to academic science has been able to keep pace as well as it has in the recent past. Most experts on fiscal and tax policy doubt that tax incentives could be designed to result in substantially increased allocation of resources to graduate education and research. In fact, the present tax system already provides many built-in mechanisms for transferring resources from the profit to the nonprofit sectors of the private economy.

Cutbacks in defense spending may produce temporary effects along the lines of 4. On the other hand, historical experience does not suggest that the demand for graduate education is very sensitive to the short-term job market. In fact, it is entirely possible that lack of posteducation opportunities may induce the opposite effect. The decline in the short-term financial advantage of going to work immediately after the baccalaureate might induce more people to continue their training, as tended to occur during the Depression. In the past the massive federal investment in research and development has scarcely influenced the fraction of college students choosing science; its effect has been mainly on the quality of the training available.

In my opinion it would be very unfortunate for U.S. science if any drastic change along the line of 2 took place. The U.S. position of world leadership in science is highly dependent on the possession of research tools with greater capability than any others in the world, and on the existence of a few outstandingly creative groups built up over a long period of time, which often set the pattern and stimulate the efforts of smaller groups throughout the country and train a disproportionate fraction of the people who become leaders and innovators in basic research in other institutions. The research-associate group in major centers often serves as the source of faculty for new centers.

Furthermore, an attempt to create new centers of excellence or achieve a wider geographical distribution of research funds primarily at the expense of existing centers of excellence would be of no service either to science or to graduate education. The inhibition of the best groups could not be compensated in leadership terms by better support of numerous other groups of a high but lesser level of competence. Inevitably it is the graduate schools of the leading institutions that set the standards to which the newer graduate schools aspire and by which they can be measured, and which must provide many of the leaders required to establish new centers of excellence. The wider diffusion of research support is an important and desirable goal, but we should not attempt to achieve it so fast that we destroy or degrade the excellence we have already achieved.

Possibly 1 may be worthier of more serious consideration. It could be achieved without as rapid an overall increase in research funds if support for nonuniversity basic research were held back, for example, in research centers. On the other hand, even here the jeopardy to our leadership position would have to be carefully considered. The principal difficulty in this area is that it is much harder to judge quality in the research centers than in the individual university research projects. Large research institutions tend to have a different social ecology than smaller university research groups. They are less individualistic, and the whole tends to be greater than the sum of the parts. At their best they provide an environment that may exploit the talents of people of average ability much more effectively than if they were entirely on their own on a university faculty. On the other hand, great laboratories tend to be evaluated by the best work that they produce, and when support is given on an institutional basis, a few excellent groups or individuals can often "front" for the whole organization, even though the total product may not be too impressive in relation to the numbers of scientists involved and the resources used. We have not yet learned how to apply the same rigorous standards to large research organizations that we do to individual research projects in universities. On the other hand, many of these organizations have a purpose other than mere excellence in basic research. It may be necessary for them to do some basic research, even if only of average quality, in order to keep a staff of the requisite level of competence to fulfill their applied mission. However, with a rising supply-demand ratio for technical people, it should no longer be necessary for such organizations to offer complete freedom of research to rather average people in order to attract them to the organization in the first place. A general tightening up in quality standards of the larger research enterprises both inside and outside universities seems both feasible and desirable in the coming decade. However, it is not clear that the real savings that might be effected in this way would be sufficient to cover the expansion required for education without substantial annual increases in the allocation of funds to academic research. In any event, very close scrutiny of major projects in space, geophysics, and other areas seems called for, not only to evaluate their intrinsic scientific merits but also to consider their impact on the rest of science. In the past, such ventures have been enthusiastically supported by the scientific community on the tacit assumption that there was no competition between these projects and the general support of little science. This assumption is valid to only a limited extent, and tends to become less so as research and development becomes a larger fraction of the national budget and of the budgets of individual agencies. This is because research budgets become more and more competitive with other activities within predetermined agency ceilings. As mentioned previously, such projects also imply commitments for operating funds merely to keep the facilities available without supporting any science.

To summarize, on the basis of educational requirements alone, it appears that a minimum annual rate of increase for university research support of 13 to 15 percent will be required for the next decade if the United States is to meet its announced goals for graduate education. This implies that by 1970 the federal money being channeled into "universities proper" should be of the order of at least $2.3 billion, of which about $1.2 billion will be for basic research. It is to be emphasized that this projection is based on very conservative assumptions regarding the development of universities in the next decade. If these assumptions do not apply, the requirements are likely to be substantially larger, and can be met only by increased research budgets or by reprogramming substantial funds from the federal support of non-education-related research. Alternatively, it is possible that the educational goals are unrealistic and should be revised downward, but this is so contrary to past cultural trends that I find it difficult to accept. One would have to demonstrate that there is some other intellectual activity that would be much more socially productive, and that would require a radically different kind of educational preparation. It should also be noted that if these goals for research support are to be met, either the budget of the National Science Foundation will have to be increased much faster than is currently envisioned (probably of the order of 30 percent a year or more) or the responsibility of the mission-oriented agencies for graduate research training as such will have to be more explicitly recognized in national budgeting.

III. CRITERIA FOR THE SUPPORT OF VARIOUS FIELDS OF BASIC RESEARCH IN UNIVERSITIES

A great deal has been written recently about criteria for support of various fields of basic science. I have already indicated that the small percentage of scientists representing the most talented and creative people should essentially be supported to do whatever they think best, within the financial limits indicated previously, since their own self-directed efforts are likely to be more useful to society than anything anybody of lesser talent could think of asking them to do. However, the people I am talking about probably represent only a small fraction (of the order of 5 percent) of those capable of doing competent and significant basic and applied research. The question of criteria, then, applies only to the activities of these less-than-top people. Even in this area it is my belief that the criteria are considerably less important than who applies them, that the fundamental problem of resource allocation within basic research is who makes the important decisions and how they are made. For example, to what extent should the cutting up of the pie among fields be left exclusively to the scientific community? At what level of detail should the financial decisions be made by the people not actually doing the work? Should resources be allocated to institutions and then divided within the institutions, or should they be allocated to broad fields and then divided within the field with the aid of representative groups of experts entirely from within the field regardless of institutional affiliation? To what degree should the system of choice be mixed — that is, with all allocation partly by institution and partly by field? If mixed, what are the proper proportions? What kind of guidelines should expert advisory committees be given? What kind of criteria, if any, apart from intrinsic scientific merit, should be used? Should the definition of intrinsic scientific merit be left implicit rather than explicit — as something that every competent scientist knows intuitively but cannot express? To what extent should judgments in special fields be left entirely to the specialists in those fields, and to what extent should the judgment of fellow scientists from neighboring fields be brought to bear?

In trying to answer these questions, I should like to try to describe an idealized resource-allocation system for basic research. In doing this I am concerned primarily with university basic research, which for purposes of this discussion, however, should include major installations outside universities, such as Brookhaven, Green Bank, or Kitt Peak, insofar as they exist primarily to serve the university community.

For the purposes of this discussion I feel that research funds should be placed in the following general classifications, which are quite separate conceptually if not organizationally:

1. The capital costs of major equipment, including in general the cost of properly housing it. By major equipment I mean the kind of equipment that would not ordinarily be provided on a research grant. The amount of money involved might vary from field to field, but I am thinking of something at least of the order of several hundred thousand dollars. In general, I have in mind really major facilities like oceanographic ships, the Mohole platform, space tracking stations, or particle accelerators. This category would include the costs of any major refurbishing or updating of such equipment.
2. That part of the operating costs of major facilities or equipment needed to make them available to the scientific community, exclusive of the cost of specific scientific work. This would include such items as ship-operating costs in oceanography, the costs of power, expendable supplies, maintenance personnel, and resident operating staff for big accelerators, the costs of computers, the logistic costs of scientific space vehicles, including the cost of procurement, launching, and tracking of a given vehicle, but exclusive of the cost of the instrument package and data analysis and interpretation.
3. The strictly scientific costs, including small permanent and expendable equipment, salaries of technical personnel, computer charges where the computer is shared by many users, publication costs, general administrative overhead, etc.

In my opinion the budgets for items 1 and 2 should be rather carefully segregated from 3. Together, they constitute what Professor Kistiakowsky has referred to as big science. The decisions regarding allocations under 1 are the only decisions regarding allocations between fields of science that should be made at the highest levels of government, for example, by the Bureau of the Budget and Congress or by the agency head. They should be made with the advice of the scientific community, but it should be recognized that they are inevitably quasi-political decisions. They are the basic investment decisions of the federal government, and they are the decisions that determine the scientific priorities for many years ahead. They are also the decisions in which the price of error is highest. In general, science-allocation decisions are less crucial because there are many investigators working independently in the same general area, and so mistakes in the decisions of one investigator tend to be compensated for by the successes of others, and the proposal-evaluation system gradually eliminates the unsuccessful ideas and investigators by a sort of free market of ideas. For the big projects involving many investigators, however, choices are much more irretrievable, and there is often no way of telling whether an alternative choice would have been better until a substantial investment has been made. For example, if the recommendations of the various panels on high-energy physics with respect to what machines should be built prove wrong, the consequence could be the loss of U.S. leadership in this field for a generation. Even if the government were prepared to retrieve the mistake by writing off the original investment and building a new machine at greatly increased expense, the time lost might be a fatal setback to U.S. leadership. For science these are the same sorts of crucial decisions as the choice of an intercontinental ballistic missile system is for national security. They are the fundamental strategic decisions of basic science, and for them criteria something like those proposed by Weinberg seem appropriate. They should be widely debated in the scientific community and elsewhere from every angle; they should ultimately be made in a highly visible and public way.

When the decisions in category 1 are made, their consequences in terms of category 2 should be clearly spelled out and understood, and should form part of the basis of the decision whether to go ahead. In projecting research budgets into the future, category 2 funds should be separately identified as such. In many cases, it would be wisest if they were not included in the ordinary individual research proposal, although this is an administrative question that may have to be decided in each individual case. It is my feeling, however, that the inclusion of fixed availability charges in individual research proposals tends greatly to confuse and complicate the proposal-evaluation process. In many cases it may be desirable to divide the availability charges between two budgets, with only a nominal charge to the individual research proposal.

Category 1 decisions also have implications for category 2 funds. It would make little sense to build facilities if support were not available for scientists to use them. On the other hand, I feel that scientific work with large facilities should not receive a specious priority just because of the political embarrassment entailed by lack of full utilization of a facility. Actually, once the commitment for the capital cost of facilities and their basic operating costs has been made, individual scientific experiments done with such facilities should compete on an equal basis in terms of scientific merit with other work that does not employ large facilities. Conversely, once the commitment to build and operate a facility has been made, I do not believe meritorious scientific work should be penalized before evaluation panels by having to bear the full category 2 costs related to the facility.

The above discussion takes care of category 1 and category 2 costs. The budgeting process should attempt to arrive at an overall government-wide level for category 3 costs in universities. This will, of course, be a sum of agency budgets, and each agency will be expected to project its category 3 costs as a budget line item. In the National Science Foundation budget, for example, this would be approximately the basic research support category, although certain of the category 2 costs of particle accelerators and oceanographic ships might be excluded and budgeted under another category, and as detailed below, certain other program costs might be included. The category 3 part of the total federal budgets — the part for university research, that is — should then be evaluated against the 15-percent-a-year growth standard mentioned earlier. I am not saying that we must have 15-percent growth every year, or that we should limit ourselves to 15-percent growth in each year. Obviously, no part of the federal budget can be sacred, and the amount of each category can be determined only in the light of the state of the economy, fiscal policy, tax revenues, and other global considerations. I am saying only that the 15-percent growth of category 3 government-wide should provide a more adequate index than we now have of how we are doing in research support. Because of the confusion of "science" with category 1 and 2 expenses, which merely build the store and keep it open but do not sell any goods, our present system of budgeting does not tell us how much science we are really buying.

At this point one must decide how to allocate the money in category 3 between disciplines and institutions. There appear to be several bases for this. Since the level of support in category 3 is being compared against a standard derived from the requirements of graduate and postdoctoral education, it ought to include not only basic research support funds but also fellowship funds, general research support funds, and some proportion of science development funds, institutional base grants, and research training grants. In other words, it ought to include the total funds being channeled into higher education by the federal government that are related primarily to research and research training, as opposed to capital investment, and to graduate and postdoctoral research training or undergraduate research activities as opposed to formal teaching activities or curriculum development.

Taking the total of category 3, we now have the question of

how it should be divided among the following categories of support:

(*a*) Project grants to individual professors or small groups of professors.

(*b*) Programmatic or coherent area grants to large groups or whole departments.

(*c*) Institutional grants, either on the basis of a formula or on the basis of specific selection criteria.

(*d*) Direct support of personnel, including graduate and post-doctoral fellowships, faculty fellowships, or career research awards, awarded on the basis of national competition.

(*e*) Direct support of personnel, but at the decision of the institution rather than on the basis of national competition between individuals in a discipline.

In this listing, the operative question is whether selection is on the basis of national competition within a discipline, or is primarily cross-disciplinary with award to institutions on either a formula or a competitive basis. (*a*), (*b*), and (*d*) are regarded as falling in the first category, (*c*) and (*e*) in the second. My own present belief is that the country has a bit overdone the matter of project support, to the point where many institutions have abnegated their responsibility for and influence over their own research activities and institutional development. Therefore, I would be inclined to recommend a gradual transition to a situation in which about 25 percent of category 3 is direct support of personnel, category (*d*); about 25 percent is institutional support, categories (*e*) and (*c*); and about 50 percent project support [including both (*a*) and (*b*)]. It seems to me the exact division between (*a*) and (*b*) is a matter for individual agency decision and negotiation with grantee institutions. It may well vary from agency to agency. As nearly as one can determine, the fiscal year 1963 figures corresponding to the recommendation above are as follows:

	Percent
Direct research and development support (including project and coherent area or program grants and contracts) . .	68
Institutional program (NSF institutional base grants and NIH general research support)	10
Training (including fellowships, training grants, career awards, and postdoctoral fellowships)	16
Construction	6

This still leaves open the question of allocation to disciplines. This presents no problem with regard to category (*c*) above, since the allocation is largely up to the institution. With respect to categories (*d*) and (*e*) I tend to be opposed, in principle, to too closely defined categorical fellowships such as those offered by the National Aeronautics and Space Administration and the Atomic Energy Commission. My observation is that students are cannier in choosing the right fields than any government administrator, and that, by and large, it is best to support the brightest people and let them choose the most promising and exciting fields, relying on the competitive salesmanship of different disciplines and the external scientific labor market to determine the actual allocation indirectly. In practice, the flexibility with which the National Aeronautics and Space Administration traineeships have been administered has apparently so far avoided what might ultimately prove to be a difficult and embarrassing problem.

If our policy is essentially to support the brightest people irrespective of field, then both government and the universities must give more attention to systematic presentation of the opportunities and promise of various fields, not only in terms of intellectual excitement but also with respect to occupational demand and social utility. I suggest that this method is superior to providing categorical fellowship support for rather narrowly defined fields. Obviously, the method of allocation on the basis of merit without reference to field is an ideal that can only be approached because of the limitations under which the mission-oriented agencies work. It might well be that some government-wide pooling of fellowship applications would be worth considering in this connection.

With respect to categories (*a*) and (*b*) there will obviously be variations from agency to agency. My feeling is that to the degree it is consistent with the agency's mission, each agency should allocate support in accordance with its estimate of the requirements of the academic community, as judged by proposal pressure and the informal advice of its program officers and consultants. Intrinsic scientific merit should be the most heavily weighted but by no means the only criterion of selection, with each agency supporting projects having a distribution of topics centered on those most closely related to its mission, but by no means confined to these. Application of this principle may actually force some gradual reallocation of resources. Of all the federal agencies, the National Science Foundation is the one that has the clearest obligation to respond primarily to the estimated needs of the academic community. There is, of course, a good deal of positive feedback between known availability of funds and proposal pressure. It is necessary to invent mechanisms to discount such effects. In this connection, widely representative advisory panels extending over several different disciplines, such as the divisional committees of the National Science Foundation or the institute advisory councils of the National Institutes of Health, must play a key role. These groups should be made more aware of the total resource-allocation problem, so that they become less inclined to promote only their own fields. Committees of the National Academy of Sciences appointed to analyze the needs of broad scientific fields and coordinated by an overall committee such as the Committee on Science and Public Policy should also play a key role in this connection. Federal agencies concerned primarily with civilian applied research should take more initiative in requesting that appropriate kinds of research be encouraged by the National Science Foundation.

There is a general problem with respect to research support

and research priorities that deserves mention at this point. One of the unfortunate side effects of the generous support of university research in the last ten years has been a tendency to denigrate the intellectual respectability of applied research. Perhaps this has always been present in the basic research community, but the size and influence of this community have reached the point where its viewpoints affect the self-image of applied scientists, engineers, and physicians, and especially the attitudes of young people toward their future careers. The generous support for academic basic research recommended in this and other papers in this series is predicated on the assumption that the healthy development of applied science and technology requires the continual infusion of people trained in basic research. Thus, to an increasing degree, many people trained in universities will be expected to move gradually into more applied areas as their careers mature. If the effect of their university training is to inculcate attitudes that make it too difficult for students to move into applied work, much of the benefit of their training will be lost to society and the justification for public support of basic research in connection with graduate education may ultimately be called into question. It is doubtful whether the long-term influence of university viewpoints on the attitudes and careers of students is as serious or as permanent as is sometimes represented. Basic research support outside universities has been increasing rapidly at a time when the supply of new Ph.D.'s was relatively constant from year to year. As a result, the opportunities for students trained in basic research to stay in basic research have been greater than ever before. This appears to be especially true in physics and biology. In chemistry, where the supply of Ph.D.'s is much larger in relation to the demand, a career in applied work is generally more acceptable. When we look toward the next decade, it appears that the situation in physics will tend to become much more like that of chemistry. I believe that the changing job market will tend to moderate the attitude of students. Still, I am in agreement with Professor Teller that there is a serious need to improve the intellectual status of applied work. This is most likely to occur when first-rate people go into applied work and provide the heroes or models that inspire youngsters. Experience shows that it is very difficult to make any intellectual activity respectable by definition, as opposed to example.

With regard to selection criteria for basic research proposals, I should like to suggest the following in approximate order of priority. Obviously, the relative importance of these criteria will vary between the National Science Foundation and the mission-oriented agencies:

1. Quality of the people proposing the research, evaluated on the basis of their past performance as judged by their professional peers and by people in adjacent disciplines. In this instance, one must be careful to avoid development of a "closed system," since those who are supported will tend to acquire a reputation that will facilitate acquisition of more support. For this reason it is particularly important that the support system provide adequately for the support of new investigators.
2. Novelty, prospects for new generalizations or important changes in outlook, and degree of penetration into important and previously unexplored territory. In this connection, emphasis must be placed on the importance of new tools. Almost every new research tool has opened up unexpected richness of phenomena. No matter how tight research budgets become, it would be dangerous to forgo the construction of really new research tools. Emphasis in research support should be on achieving new understanding or generalizations, and not merely the assembling of new data for their own sake. Measurements should be informed by hypotheses or expectations.
3. Relevance to recognized practical problems, assuming there

is a reasonable prospect of progress. This criterion must be applied with caution and good judgment. Applied too narrowly and unimaginatively it can result in the support of rather trivial and pedestrian research. There is always a tendency to support applied research projects that are really basic research, but whose intrinsic scientific merit does not make them competitive with other basic research proposals. To the degree that relevance to practical problems is claimed as a basis of support, certain hard questions should be asked. What is a solution to the problem worth? How critical is a particular piece of information to the solution? What is the probability of success? What is the probability of unanticipated development or by-products? In answering these questions, the advice of people with experience in the practical problems involved should be sought, as well as the advice of people concerned only with the intrinsic scientific merit of the work.

4. Educational value, in both the strict sense and in the broader sense of extending the capabilities of bright people or groups of people. Will the research tend to stretch the limits of an existing technology that is likely to have other applications? Will it exploit a new technology not previously available as a basic research tool? Will it help maintain a standby capability in terms of people whose activities may become nationally critical in the future, as in the nuclear weapons laboratories? Will it help train graduate students, and enhance our resources for graduate education?

The preceding discussion has been concerned primarily with the criteria that should be used in allocating resources to basic and applied research in universities as well as to fellowships and other forms of support that indirectly subsidize research. The criteria suggested apply not only to what Dr. Kistiakowsky discussed as little science, but also to big science insofar as it is primarily connected to universities and graduate education. In the area of academic research the emphasis is on the autonomy of science and on primarily scientific criteria of choice, although certainly other considerations such as potential relevance to the mission of the supporting agency must be given significant weight. However, it must be recalled that less than 50 percent of all the basic research supported by the federal government is conducted in universities proper. We must now discuss the criteria for support of basic research that is not connected with graduate education.

IV. INSTITUTIONAL RESEARCH

The term "institutional research" is designed to cover a broad spectrum of activities ranging from university-based research institutes to industrial laboratories. Basically it is characterized by the fact that the great majority of the scientists are full-time career research workers not engaged in classroom teaching. As pointed out above, the social ecology of these institutions differs from that of universities proper, and judgments concerning their support should be based on different criteria. Institutes of this sort can be further subdivided into two types: (1) those primarily concerned with basic research, having the aim of advancing some generally defined broad area of scientific knowledge, or perhaps a group of such areas, but usually connected by some common theme or object of study, and (2) those primarily concerned with an applied objective usually related to the mission or missions of some federal agency.

Sometimes a single laboratory may combine both functions in some degree; for our purposes it should then be considered as two separate institutions. The great national and government laboratories usually fall in category 2, as do industrial laboratories. The only exceptions are laboratories like Brookhaven, the Green Bank Radio Astronomy Observatory, or the Kitt Peak

Observatory. These are really extensions of university research. They have a service function in relation to universities, but their career research staffs are independent scientists in their own right. In a sense, however, they still serve an instructional function in that they help train graduate students and faculty members and postdoctoral associates in the newest techniques of their science. They can also undertake research problems demanding greater continuity and cooperative effort than is possible in a university department with other responsibilities. The basic research laboratories in category 1 should not be judged by the same criteria as those used in connection with universities. In the first place, greater scientific productivity should be expected of such groups, since they do not have other responsibilities. In the second place, they should truly serve their function of supplementing and assisting the universities; the resident staffs should not be so large as to pre-empt the facilities for their own experiments. It seems to me that in periods of limited research funds the expansion of such institutes should have lower priority than the expansion of university-based research, which is more closely related to teaching. The local management should be given great freedom and should be promised continuity of support but not necessarily continually expanding support. The creation and support of such institutes ought to be based on general criteria for the support of various fields of "big" science along the lines suggested by Dr. Weinberg. In terms of quality, such institutions ought to be subjected to standards similar to those applied to university groups. Insofar as they carry out independent basic research, such institutions ought to concentrate on types of research requiring special facilities, an unusually programmatic or long-term type of approach involving the closely coordinated activities of many senior scientists, or other basic research activities that are unsuitable for the individualistic style of university research. Conversely, universities should concentrate on types of research that lend themselves to the individualistic approach.

Most of the great national laboratories fall in category 2, that is, they have an applied mission. It is entirely right and proper that such laboratories should do a substantial amount of basic research, since experience shows that participation in basic research enables them to attract better people, to keep their staffs alert to new scientific developments of potential importance to their missions, and generally to perform better. However, the total support for such establishments should be based on the national importance of their applied missions and on their long-run success in performance. The fraction of support that goes into basic research should be largely a local management decision. Such a laboratory should not receive increased support for basic research purely on the basis of the excellence of its scientific work or the number of papers published by its staff in reputable scientific journals. These may be indications of the general quality of the laboratory, but are not enough by themselves to justify its support. If this policy is followed, increased support for freewheeling research activities should be provided essentially as a reward for success in the performance of the applied mission, thus serving to give the whole staff a stake in the applied goals of the organization rather than setting up a status system in the laboratory that isolates the basic research from the rest of the laboratory. The so-called "independent research" supported by several agencies as part of the overhead on procurement contracts with profit organizations contains such a built-in incentive for success in its applied objectives, and a somewhat similar incentive system might be encouraged with respect to nonprofit institutions doing applied work.

A special problem has arisen in connection with support of basic research by the federal government in industrial laboratories. Not only do many agencies support project contracts with industrial laboratories on a somewhat similar basis as that applied to project grants to university groups, but whole laboratories exist primarily by performing research services for federal agencies. Some of this project activity represents excellent scientific work. On the other hand, there is a real question in my mind whether the basic research project contract is the proper mechanism for supporting industrial groups. This is especially true when research proposals from these groups are evaluated primarily on the basis of intrinsic scientific interest or merit rather than on the basis of their potential contribution to a specific applied objective. It is hard to lay down hard-and-fast rules in this matter, but in general, it is my opinion that institutional-type support is preferable for industrial groups. In this type of support the basic research is supported by the local management as part of a general program aimed at an applied objective. Government laboratories and federally supported research centers also occasionally attempt to supplement the support from their parent agencies by seeking basic research contracts with other government agencies in competition with academic research groups. In principle, this is undesirable; I would be strongly opposed, however, to blanket rules or regulations concerning it, and it would be unwise to alter abruptly the system of support that has grown up over the years. Such a sudden change would be unnecessarily disruptive. However, this is a general area that Congress may wish to examine, and agencies now supporting industrial and federal laboratories under small project grants and contracts should be encouraged to devise new support mechanisms more consonant with the institutional character of these organizations. The extent of this type of project support is not known at present, but it has an open-ended character that could make it a potential drain on tight basic research budgets if it were not carefully watched.

Occasionally, it is advantageous for agencies to make contracts with industrial organizations with a view to exploiting unique industrial skills in getting rather specific jobs done, usually in relation to some broader applied program or to provide needed tools or materials for basic research. Examples might be the growing of crystals for experimental purposes or the development of new research equipment for which the potential market may be insufficient to justify private financing of the development costs. I have no criticism of contracts of this type.

V. CONCLUDING COMMENTS

The basic thesis of this paper adds up to the conclusion that the concept of a total science budget, which is implied by the questions asked by the House committee, is probably not a very meaningful or significant one. Only in the restricted area of academic basic research does the concept of a government-wide "science budget" make a certain amount of sense. Even here it is essential to separate out the costs of major equipment, both the capital costs and the cost of keeping it available for the use of the scientific community. The rest of the "science budget" ought to be considered in a different context, in which the value of research and development is judged in competition with alternative means of achieving the same objectives. In these areas I think that Congress and the Administration ought to consider primarily the total resources that it is worthwhile to devote to a general objective, and then regard as tactical rather than strategic the decision as to what fraction of these resources should go into research and development. Inevitably, such decisions are quasi-political and must be settled by debate among the various groups concerned; the voice of the scientists should be heard but should not be conclusive in this part of the debate. Basic research outside universities — more than 50 percent of the total — should be judged in terms of its potential contribution to the missions of specific agencies.

SEVEN □ The Future Growth of Academic Research: Criteria and Needs

The following chapter is a paper prepared for a seminar held by the Brookings Institution in Washington during the late fall of 1965. It is essentially an elaboration of some of the themes dealt with in the preceding paper for the report "Basic Research and National Goals."

INTRODUCTION

This paper will be concerned with how to determine our national investment in research and its rate of growth in the future. This is the subject which has come to be referred to as "science policy." It is only recently that public policy has become concerned with science as such. Up until the last five years federal support of science has been determined almost entirely as an incidental by-product of other federal missions such as health, national defense, and agriculture. These missions have been expressed organizationally as federal agencies and departments and their expenditures for research have been largely determined in competition with other activities and functions within the agency. Since the end of World War II explicit "planning" for science as a whole has been considered of relatively little importance because the overall rate of expenditures for scientific activities has been growing sufficiently rapidly to permit a large margin for new enterprises and the entry of new scientific investigators into the system. Nevertheless, as the totality of research and development activities has grown, there has been an increasing demand for integrated planning even in basic research. Such a demand for planning tends to be naturally repugnant to basic scientists, who look upon the scientific endeavor as a sort of free enterprise system of ideas. Nevertheless, the leveling off of the growth of research budgets since 1963 has convinced even the scientific community that a more systematic confrontation of alternatives will probably be necessary even within basic science.

There is increasing public and political acceptance of the fact that basic science cannot be planned in detail, except possibly for the construction of very large and expensive research tools such as space vehicles or accelerators. Scientific planning is concerned not with the central direction of science but with planning and projecting the resources for scientific research and technology. The dialogue between the scientific and political communities is concerned with just what are appropriate areas for planning. What are the alternatives to be confronted? Planning is inevitably concerned with aggregates, usually expressed as financial obligations for support of research, which are the summation of many diverse individual activities having some property in common. The politician and the administrator are often bewildered by the categorizations of scientific activity, which are multidimensional in character. Yet it is precisely the appropriateness of the various possible characterizations of aggregates which lies at the heart of the planning problem and the determination of criteria for choice. Of course, this multidimensional character is not entirely unique to science. There are few federal activities which can be defined by a single purpose, and once an activity serves many purposes, one is faced with the problem of how to combine the different purposes into a single objective function. For in the last analysis choices are unidimensional; the ordering of alternatives is a scalar function.

In the U.S. system, programming for scientific activities really begins, or should begin, more at the bottom than at the top. One does not start with an *a priori* allocation to broad categories of scientific activity and then suballocate to projects within science. Instead, specific projects originate at many levels within the federal and nonfederal structure and are gradually selected as they move up through the decision-making process. There is no government-wide allocation for basic research, or for physics or chemistry as such. Instead, each of the federal agencies tends to balance all of its activities within its own budgetary ceiling. Of course, the ceilings assigned to agencies are influenced by their scientific responsibilities as well as their other missions. The budget of the Atomic Energy Commission, for example, is not determined independently of the fact that it has a responsibility for high-energy physics, but an increase in its high-energy physics expenditures does put pressure on all its other activities.

Several factors, however, have led to increased scrutiny of research and development activities as an aggregate. The most important of these is the sheer magnitude of the total — of the same order as total sales of the automobile industry, for example. A second factor is the increasing budgetary importance of independent agencies whose mission is mainly defined in terms of science itself — notably the space agency, the National Science Foundation, and, to a lesser extent, the Atomic Energy Commission. A third factor is the growing public attention to higher education, and belief in the close connection between academic science, graduate education, and regional development.

ALTERNATIVE CATEGORIZATIONS FOR SCIENCE

If planning for the totality of national scientific and technical resources is to proceed on a more rational basis than at present, the first step must be a better understanding of the various alternative categorizations of science and its implications. Whereas development, which constitutes about two thirds of federal support of technical activities, tends to be concentrated in a relatively small number of large projects, research, with which I am primarily concerned in this paper, is distributed among tens of thousands of individual projects. To collect these projects or expenditures into aggregates only has meaning for planning to the extent that the individual components have some significant property in common which is related to a national objective. Otherwise the categorization tends merely to confuse discussion. For example, it is not at all clear that "basic research" by itself has any significance as a category when it includes activities ranging all the way from a pure mathematician proving a theorem in topology to a huge engineering and logistic effort such as a space probe to Mars or a Mohole project. Yet these disparate activities are actually included in the category basic research by the National Science Foundation in its reporting of federal R and D expenditures. I am not questioning the validity of the categorization but only its significance for decision-making. Is there any proper sense in which such disparate activities in basic research should be regarded as competitive, or is the Mars space probe more properly competitive with, for example, the national highway program, the maritime subsidy, or the supersonic transport?

In the discussion of scientific activities a number of dimensions of classification have been used. These may be summarized as follows:

1. The degree of fundamentality or applicability, for example, basic research, applied research and development.
2. The scientific discipline, for example, physics, chemistry, or biology.
3. The social purpose or function of the research, for example, health, defense, natural resources.
4. The institutional character of the research, for example, academic, research institute, industrial, government laboratory.
5. The scale or style of the research, for example, "big" science versus "little" science.
6. The object of study; for example, oceans, atmosphere, or space. Although subjects such as oceanography, meteorology, and space science are often confused with scientific disciplines in common parlance, they are actually multidisciplinary studies applied to a single object or aspect of the environment.

Since these are different ways of classifying the same aggregate they overlap to a large extent, and different classification schemes are more appropriate for some purposes than others. In general, for example, the more basic or fundamental a research activity is, the more appropriate is a subcategorization by scientific discipline. Similarly, a disciplinary classification is more appropriate for academic research than for industrial research or research in government laboratories.

It is also important to understand the limitations and ambiguities in each of the categorizations listed above. In the basic-applied spectrum, for example, the same research project may be looked upon as applied by one investigator and basic by another, depending upon the particular institutional environment in which he is working. Two research projects which start from the same point and ask the same questions may develop along entirely different lines owing to the attitude of the researcher and the nature of the scientific communication network within which he works. Furthermore, the basic-applied classification is constantly changing. Last year's basic research is often this year's applied research. Not only are applications found for new discoveries, but new applications often generate new questions for basic research or new technologies which can be applied as tools in basic investigations. The basic research of one field may be the technology of another. Whether research is regarded as basic or applied depends upon the time horizon within which one looks at it. It has been argued with some justification that all fundamental research in biology and biochemistry should be regarded as applied. Our understanding of fundamental life processes is still so rudimentary that almost any advance in understanding is likely to find applications rather quickly in medicine or agriculture. One of the least applicable of all branches of physics, namely elementary-particle physics, nevertheless places extreme demands on advanced technology. Thus many of the techniques developed as an incidental by-product of elementary-particle research have had an important impact on technology and on other fields of science even though the concepts and theories of the field itself are little related to presently foreseeable applications.

The disciplinary categorization is equally limited. This has become increasingly true as advances in understanding have permitted the use of the concepts and techniques of the physical sciences in other disciplines. The growth of science in the postwar era has been characterized most by the spectacular growth of hybrid disciplines such as geophysics, geochemistry, biochemistry, chemical physics, computer science, systems analysis. Techniques such as radiocarbon dating, radioactive tracers, paper and gas chromatography, microwave and nuclear resonance spectroscopy, and X-ray diffraction have spread rapidly into all fields of science. Interdisciplinary subjects such as oceanography, atmospheric sciences, and space science draw on all the more classical disciplines, and it is difficult to tell at what point they do or should become disciplines in their own right. Whole areas of research often move from one field to another. For example, atomic spectroscopy, which used to be a major branch of physics, has now moved almost entirely into astronomy. Similarly, molecular spectroscopy has largely moved from physics into chemistry. Cosmic rays have largely moved from physics into a branch of space science. The theory of low-energy nuclear reactions has become an important branch of astrophysics. It becomes increasingly difficult to define a discipline except by the organizational framework within which it is pursued; for example, physics is what is done currently in academic physics departments.

The classification of research by social function encounters similar difficulties. Missions such as defense and space are so broad that they require a large amount of general purpose research which is equally relevant to other missions. Both, for example, are heavily dependent on electronics, which in turn depends increasingly on advances in solid-state physics and a number of basic engineering disciplines such as information theory. Yet unless this general purpose research is carried out in the environment of the purposes of defense or space, it may not find application to these missions. If we define as defense research any research supported by the Defense Department or its agencies, then we encounter the problem that the volume of general purpose research in a given area which should be supported by the Department of Defense is not independent of the volume of similar research which is being supported by other agencies, including the National Science Foundation. On the other hand, no special purpose agency can afford to depend for its general purpose research entirely on other agencies, since each area of application requires a slightly different emphasis. Such general purpose research blends continuously into the more specifically mission-relevant activities of the agency, and no simple rule can be used to draw the line between mission relevance and general purpose. Furthermore, there are several areas of importance which cut across the social purposes defined by present federal organization. They are vital to the mission of a given agency, but also transcend the mission of any one agency and constitute a new social mission. One of the best examples is the subject of oceanography. Another is the whole subject of information technology.

Furthermore, there is the whole issue of how much the support of disciplines as parts of missions should be "counted" in evaluating the total activity in a given discipline. This difficulty was encountered in acute form by the Physics Survey Committee of the National Academy of Sciences in its attempt to arrive at an estimate of federal expenditures for physics and astronomy.[1] It found that gross reclassifications of NASA ex-

[1] "Physics: Survey and Outlook, a Report on the Present State of U.S. Physics and its Requirements for Future Growth by the Physics Survey Committee of NAS–NRC" (Washington, D.C., 1966); see Chapter 6, pp. 81–90, especially figures 4 and 5.

penditures resulted in radical changes in the total picture for physics and astronomy. Similar questions arose concerning National Institutes of Health support of chemistry[2] in connection with the work of the Committee on the Survey of Chemistry of the National Academy of Sciences/National Research Council. This type of difficulty has arisen in almost every field where an important mission of a federal agency has been extensively dependent on a particular scientific discipline, and yet where the overall organization is not explicitly discipline-oriented.

A great deal of attention has been given recently to the classification of science into "big" and "little." [3] A principal difficulty with such a classification lies in the fact that there is actually a continuous spectrum of activities between big and little science ranging from a scientific space probe to an individual professor in his laboratory working with two or three graduate students. There is clearly a strong argument for some segregation of the logistic costs of types of scientific research requiring expensive equipment or large-scale institutionalization. Otherwise, there is a serious danger that such support costs will consume a disproportionate share of the total funds available for support of science. This has apparently happened, at least for short periods, in the case of nuclear-structure physics and oceanography.[4] Nevertheless, it is also true that many of the specific scientific activities carried on in big science have some of the characteristics of little science and some of the same educational functions. There are real dangers both to

[2] "Chemistry: Opportunities and Needs, a Report on Basic Research in U.S. Chemistry by the Committee for the Survey of Chemistry, NAS–NRC" (Washington, D.C., 1965); see chapter XII, pp. 163–181, especially p. 170.

[3] "Basic Research and National Goals," National Academy of Sciences, Report to the Committee on Science and Astronautics, U.S. House of Representatives (March 1965); see especially essays by G. B. Kistiakowsky and Harvey Brooks.

[4] *Ibid.*, p. 81.

the health of the field and to the health of graduate education in treating big science as too "different" from the rest of science and especially that part in which graduate students and professors are involved.

INSTITUTIONAL CLASSIFICATION OF RESEARCH

After much thought about this problem I have concluded that the most convenient and meaningful categorization of research activities is probably in terms of the primary purpose of institutions or organizational subdivisions which perform it. This has several things to recommend it.

1. It tends to be much more objectively ascertainable than categories such as basic and applied or classifications by social purpose.
2. The contribution of research to society is probably most strongly conditioned by the environment in which it is performed, that is by the communications network within which it takes place and the pattern of movement of people between different types of activity.
3. Other types of categorization are easier to use, once the institutional framework is defined. For example, within the academic context alone classification of research by scientific disciplines is much more meaningful than in industrial or government laboratories. Similarly in governmental laboratories or federal contract research centers classification of research by social purpose tends to be much more meaningful than in universities.
4. The institutional framework considerably clarifies the relationships between big and little science. In general big science tends to be carried on in institutions especially created for the purpose, whereas little science tends to be a primary function of the regular university structure. Even within the university separate organizational struc-

tures tend to be created to carry on big science, and on the other hand, within government or federal laboratories, although there may be a large volume of little science activity, it does not generally constitute the *raison d'être* of the institution and is not an end in itself as it is within the university structure.

5. The institutional framework also clarifies the relationship between basic and applied research. As I have already mentioned, the line between basic and applied research tends to be determined by the environment in which it is conducted. Thus what may be regarded as basic in an academic context might be regarded as applied in an industrial one. In short, the categories such as basic and applied, physics versus chemistry, big science versus little science, tend to be different in different institutional contexts.

For the purpose of discussing the resource allocation problem for scientific research, I have found the following categorizations most useful:

A. Nonacademic fundamental research. This is basic and very fundamental applied research performed in institutions whose mission is defined in terms of nonscientific objectives such as health, standards, defense, agriculture, and so on. These are functional missions, where the term functional may denote either a specific end product or service or simply the cultivation of a general area of technology related to the overall mission capability of the parent agency. This includes all industrial basic research, plus basic research conducted in most government laboratories, in federal contract laboratories, and in nonprofit technological institutes.

B. National center basic research. This research, like that in category A, is primarily institutional in character, but where the mission of the performing institution is defined in terms of an area of science rather than in terms of technology or a socially desirable objective. In most cases, national center research is clearly related to and supportive of university research, and may be regarded as an extension of academic research. Some national centers, such as the Cambridge Electron Accelerator or the Lawrence Radiation Laboratory at Berkeley, are actually parts of universities. Brookhaven, the National Radio Astronomy Observatory, the Kitt Peak National Observatory, the National Center for Atmospheric Research, and the Woods Hole Oceanographic Institution are all examples of national centers. On the other hand, the Lincoln Laboratory, the National Bureau of Standards, the Naval Research Laboratory, the Livermore Radiation Laboratory all belong to category A because they are primarily technology- or mission-oriented rather than science- or discipline-oriented. National center research usually involves a component of "big science," but may involve only a "supercritical" interdisciplinary organization of scientists. Such research is characterized by a larger supporting engineering and service effort than is characteristic of normal academic research, and hence a higher ratio of supporting personnel to independent research investigators, often including professionals of training and experience equivalent to that of the independent scientists.

C. Academic research, including both basic and applied research. This is research conducted within the framework of graduate education and faculty activity, most often within academic departments organized by scientific disciplines. Characteristically it is "little science" with a low ratio of supporting personnel to independent investigators and apprentice researchers. One may here distinguish between two subtypes of academic research according to its relationship to the external sponsor, as follows:

1. Mission-oriented academic research, which is supported primarily for the sake of research results of some interest to a nonscientific mission or purpose, or for the sake of

training people who may ultimately contribute to such a mission. This includes essentially all sponsored research other than that supported by NSF, although it may include certain basic research activities, such as elementary-particle physics in AEC, which a mission-oriented agency may have undertaken as a national responsibility beyond what was strictly justified by its assigned social mission.

2. Non-mission-oriented academic research — that is, research in which the education of scientists and adding to the store of human knowledge with only secondary consideration of possible relevance to potential applications are the primary motivations. Most of the basic research supported by NSF in universities comes in this category, even in engineering.*

For most purposes, the term "academic research" is congruent with the category "research in universities proper" used by NSF in reporting on federal expenditures.[5]

It must be recognized that the above classification scheme, like any other, has serious limitations, and its value will be

* The term "basic research in engineering" causes a great deal of confusion, since engineering is ordinarily thought of as applied by definition. Nevertheless, there are a number of basic scientific disciplines which have become traditionally associated with engineering in universities even though they are pursued largely for their own sake, and are just as fundamental as the disciplines associated with the natural sciences. The basic sciences associated with engineering are primarily those concerned with the behavior of manmade systems—information theory, the theory of structures, the theory of feedback and control systems, computer and systems theory—but they are nonetheless fundamental. Also there are subjects such as fluid mechanics, solid mechanics, and thermodynamics—what might be called macroscopic or classical physics—which have been largely taken over as basic engineering sciences. All these engineering sciences are distinct from engineering as the art of applying the mathematical and physical sciences to the meeting of human needs.

[5] "Federal Funds for Research, Development, and Other Scientific Activities," Vol. XIII and XIV, NSF 65–13, 14 in "Surveys of Science Resource Series"; for general definitions including classes of performers see pp. 92–96 of Vol. XIII; for list of Federal Contract Research Centers see pp. 104–105.

strongly dependent on the purpose for which it is to be used. In this paper I will be primarily concerned with criteria for the growth of academic research in the aggregate, and my general thesis will be that academic research is the only research support *aggregate* for which it is desirable to have some policy. Other aggregates should emerge only as by-products of decisions regarding the creation of specific institutions and capital facilities, as outlined in my paper in "Basic Research and National Goals." (See reference 3.) In other words, academic research is the only aggregate which should be considered, as an aggregate, as competitive with other federal expenditures. Other federal expenditures for science and technology should be considered primarily on the merits of the individual activities in competition with nonscientific expenditures, not on an aggregate basis.

One difficulty in this classification scheme arises from the activities in national centers which provide considerable support for graduate students and faculties. In the various disciplinary studies by the National Academy, national center research tended to be handled differently by different groups. In its assessment of the current level of support for physics, for example, the Physics Survey Committee counted 80 percent of the elementary-particle research in national centers such as Brookhaven as academic research.[6] This had the effect of nearly doubling the amount of physics research as compared with the NSF category of "universities proper." The justification for this assignment was that 80 percent of the machine time in the national centers is devoted to experiments conceived by and carried out in collaboration with faculty members and graduate students, and only 20 percent of the time is controlled primarily by resident research staff. A similar situation pertained to research in optical and radio astronomy.[7] On the

[6] Reference 1, see footnote b under Table 14, p. 104.
[7] Reference 1, pp. 91–92.

other hand, the Chemistry Panel excluded all but about 20 percent of the chemical research undertaken in such university-based organizations as the Berkeley Radiation Laboratory and the Atomic Energy Laboratory at Ames and included as academic research none of the basic chemistry being done in such centers as Brookhaven, Argonne, or Oak Ridge. The figure for Ames and Berkeley was arrived at by assigning the same cost per graduate student and faculty member to these laboratories as the average for research in the chemistry department of the same university. The chemistry report excluded entirely the cost of operation of AEC reactors and accelerators which are extensively used by university chemists in universities for research in nuclear chemistry. Indeed, for the most part, the machine costs for nuclear chemistry research are borne by research in nuclear-structure physics or elementary-particle physics.[8] A good case can be made for the procedure followed by either the Chemistry or the Physics Panel, but the differences in procedure make it necessary to be extremely cautious in drawing conclusions regarding the comparative levels of support for university chemistry and physics. A further complication is that whereas the Chemistry Panel defined chemistry support as only that going to university departments of chemistry and excluded biochemistry departments and basic science departments in medical schools, the Physics Panel included a number of activities supported in other than physics departments, including both engineering and chemistry. It is doubtful whether these differences would have a major effect on totals, but they further contribute to uncertainties in the comparison between fields.[9]

Another problem with aggregating research support arises from restriction of most statistics to what the Chemistry Panel has termed "explicit support." There is also a large component

[8] Reference 2, see footnote b of Table F1, p. 217, also p. 171.
[9] Reference 1, p. 103; reference 2, p. 151.

of what the same panel termed "hidden support," which includes faculty salaries paid out of university funds for research time, fellowships, amortization of faculties, unreimbursed indirect costs, and so on. For chemistry the panel concluded that explicit federal support constituted only 55 percent of the total funds going into academic chemistry research.[10] Hardly any information of this sort is available for other fields of science, and yet if aggregates for academic research are to be used as a basis for science policy, it is clear that such statistics will be essential.

The remainder of the paper will be concerned primarily with academic research, and with such research outside universities as can be categorized as academically related — mainly, research in national centers. I shall not be further concerned with research in mission-oriented institutions where the mission is other than science itself or scientific education.

THE FUNCTION OF THE UNIVERSITY IN AMERICAN SOCIETY

It is impossible to discuss the needs of academic research support without some assumptions as to the purposes and functions of the modern university in American society. During the last fifty years there has been a dramatic change in what society expects of universities. From institutions whose function was primarily the teaching of a fixed body of knowledge to students, they have been transformed into "multiversities," expected to be omnicompetent. Public and political attitudes have not uniformly caught up to this transformation, nor have the universities themselves always been able to define their own functions and recognize their own limitations. There is no one organization or structure of internal communications which

[10] Reference 2, pp. 169–171.

can serve all the purposes for which scientific work is performed. The very characteristics which make universities effective performers of most types of basic science often make them rather ineffective in the application of science to well-defined short-term goals.

I would list the primary functions of the university as follows:

1. To add through research and scholarship to man's understanding of himself and the world in which he lives.
2. To integrate newly acquired knowledge into the total intellectual structure, to systematize and organize knowledge for each generation.
3. To communicate existing understanding and knowledge through formal teaching, writing, and other forms of communication to the intellectual public, and to train the next generation of scholars through apprenticeship.
4. To be the custodian of the intellectual standards of society, and to maintain intellectual leadership in the major fields of human knowledge and its long-term applications.

It is fairly clear from the statement of these functions that research is essential to all of them, but especially to the maintenance of intellectual leadership. It is essential that the universities be able to attract and hold a fair proportion of the best minds in each generation. Universities should be concerned with the application of knowledge only insofar as it enriches the total intellectual enterprise and contributes to the proper training of future leaders in the professions. Because of this their natural organization is along disciplinary lines. The very word "discipline" carries the connotation of standards. A discipline is a certain mode of approach to the acquisition of knowledge and understanding. By its very nature it is partial; to some extent it sacrifices breadth to depth. This is not to say that interdisciplinary activities are not important, but they should never supersede the disciplines themselves. In fact, interdisciplinary activities are important in universities precisely because they provide the seed from which *new* disciplines are ultimately generated, and hence some measure of interdisciplinary activity is vital. It also performs an important link to the needs of society, especially in connection with the training of people for the learned professions which are concerned with the application of the fruits of learning to the needs of society.

Universities are characteristically individualistic, and they are at their best when they try to exploit the freedom of the individual intellect. Separate institutions or "establishments" are best for programmed research and for a well-coordinated multidisciplinary approach to problem areas, as opposed to disciplines. The process of transfer of research results to technology or application naturally acquires more prestige in nonacademic institutions. Academic institutions have done well in pioneering certain technologies such as computers in their early stages, when the primary requirement was for new ideas and concepts and the work could be done by small groups. But for concrete applications and systems universities lack the continuity of effort, and the organizational discipline necessary to channel the effort of many people toward a planned large-scale goal. There is no proper place in their social structure for the necessary supporting engineering or service efforts. Historically they have occasionally been able to accommodate their structure in an *ad hoc* fashion to large-scale efforts in "big science" when the intellectual incentives were strong enough, but this is not natural, and generally involves certain internal social strains. Examples of the social problem are the difficulties in attracting academic chemistry faculty into a supporting role in solid-state physics efforts, the difficulty in attracting physicists and engineers into problems of medical research and instrumentation, and the lack of interest in engineering faculties and students in the design and construction of complex research instrumentation for other academic disciplines, despite the fact that this would appear to be a logical activity for engineers within a university.

Nevertheless, the types of research which go on under categories A, B, and C above are by no means mutually exclusive. Nonacademic institutions and national centers may be expected to carry on considerable research which is indistinguishable from the academic variety, but their reason for existence disappears if *all* their activity is of this type. Similarly research which is very specifically applied, consisting of short-range process or product improvement, or service to a particular customer, or which requires extensive employment or permanent research or supporting personnel, or a high degree of planning and programming, may be done occasionally in universities as an expedient but is inappropriate if it becomes a significant fraction of the totality of university research.

BASIS OF POLICY FOR ACADEMIC RESEARCH

The following are the general guiding principles which in my opinion should govern policy toward support of academic research in the next several years. Obviously the actual allocation of funds will depend upon total federal budgets. In this section we deal with the qualitative considerations which are relevant to policy, and in the following section we deal with the quantitative basis of calculation.

1. At their best, good teaching and good research are inseparable. Each should reinforce the other. Research activity is needed to keep faculty intellectually alive, but the continual effort to synthesize and present existing understanding to less specialized students helps the faculty member to illuminate his own research activities and open up new lines of inquiry. Furthermore, students are exposed to a variety of specialists, ideally enabling each generation of students to stand intellectually on the shoulders of several of their predecessors with different specialties. Thus, for example, today's graduate student in physics or electrical engineering should learn more abstract mathematics than

is known to his physics or engineering professors, while the student of biology should learn more physics and chemistry than most of his biology professors know. It is partly by this synthesis of many specialties in the minds of the new generation of apprentice researchers that each new generation is able to penetrate more deeply than its teachers.

2. It should be national policy to help create an environment in the universities where some of the very best minds of each generation are attracted to remain in academic life. No significant intellectual discipline or line of basic research should remain unrepresented in the universities, and in general, universities should be encouraged to maintain the balance of intellectual leadership in all important fields. Leadership does not necessarily imply a numerical preponderance of research workers or of financial support, but it does imply enough support to maintain some university groups at the cutting edge of every major field. It implies that when national decisions are taken to initiate new lines of research or expand existing efforts, priority consideration should be given to universities, and new efforts should be undertaken outside universities only if, for reasons outlined in the preceding section, other types of institutions are better fitted to carry them out. In other words, the burden of proof should lie with new nonacademic research undertakings to justify themselves as compared with enhanced academic support.

3. It is now a goal of national policy for education to make available educational opportunities, up to the maximum they are capable of absorbing, to all American youth, regardless of their ability to pay their own way. This is not only a matter of social justice, but a matter of sound social investment policy from the standpoint of economic growth and a better society, since it ensures selection of the ablest people from the largest possible pool of students.

The trend line for Ph.D. production in science and engineering has been rising at about 6 percent a year since 1900. There is no indication of or reason for a downward departure from this trend in the next fifteen years, and the current trend is over 10 percent a year and is expected to continue for several years. Academic research support should be geared to ensure availability of a quality educational experience within this trend.

It is important to remember that this upward trend in Ph.D. production is not unique to science and engineering. In fact, the distribution of Ph.D.'s between the major areas of learning has remained surprisingly constant over a period of sixty years despite structural changes in American society.*

The sociology of the educational system will tend to change in such a way that we will be able to maintain approximately the present degree of selectivity among students and research projects as the size of the effort grows. This assumption is based on evidence that a large proportion of students at the school level are far from realizing their full educational potential due to economic, class, and geographical disabilities. Maintenance of selectivity obviously depends on accelerated efforts in curriculum develop-

*The percentage of awards in science and engineering was 54.9 percent in 1964 and has varied by less than 1 percent since 1900. The number of bachelor's degrees in science and engineering has remained similarly constant at 24.5 percent of the total since 1920. Between 1950 and 1963 scientists and engineers increased in numbers by 91 percent as compared with an 85 percent increase for "professional, technical, and kindred" workers of all types. Scientists and engineers constituted 16.5 percent of the professional work force in 1963. The annual growth in the professional work force is about 4.8 percent as compared with 1.2 percent for the labor force as a whole. Even in the next fifteen years the professional category would grow only from 10.4 percent of the total labor force to about 17.7 percent of the total, while scientists and engineers would grow from about 1.7 percent to 3 percent of the total labor force. While it is obvious that the fraction of technical people will have to level off before the end of the century, such a leveling does not appear to be imminent and would probably not affect the projections being discussed in this paper.

ment and science education at all levels, not just on academic research.

4. The rate of incorporation of technical people into our society is limited primarily by supply rather than demand, and this will continue to be so. In other words, scientific and engineering manpower availability acts as a spur to development, rather than the economy exerting a pull on the manpower supply. This is true on a long-term basis despite short-time fluctuations in supply and demand. The supply tends to generate its own demand through opening up new opportunities. Evidence for this includes the continuing immigration of foreign scientists and professionals to the United States.
5. A pluralistic system of decision-making for academic research will optimize the distribution of funds. Each organizational subunit in the system should control some funds about which it can make significant choices; for example, a plurality of federal agencies with different missions, panels of scientific peers, institutions, fellowship panels, individual researchers. Policy should also be designed to encourage the expansion of private and nonfederal support to match approximately the expansion of federal support. This appears to have happened in practice in the past, although federal research support has expanded somewhat faster than national expenditures for higher education from all sources.
6. It should not be expected that everyone trained in basic research will be encouraged to stay in basic research. On the contrary, the output of Ph.D.'s in a discipline should normally be expected to exceed the needs of that discipline within the universities. Part of the social value of basic science training stems from the migration of scientists into other fields and into applied science and technology.

A study of a sample of 1955–1960 Ph.D.'s from the National Register indicates that in the physical sciences only 35 percent were employed in colleges and universities by 1962, while 45 percent were in government and industry. Only 30 percent of the 1960 cohort of chemists remained in colleges and universities.[11] In the life sciences nearly 60 percent remained in colleges and universities, but this is explained by the fact that, through medical and agricultural schools, the universities are much more heavily engaged in the application of the life sciences than of the physical sciences. The studies of the Physics Survey Committee indicate that each of the major subfields of physics studied absorbs fewer Ph.D.'s than it produces. I believe this is a healthy situation. On the other hand, it would be unhealthy if there were too few outlets within universities for the Ph.D.'s trained in any field over a long period.

7. Policy should aim to diffuse the geographic and institutional distribution of basic research capability, but should not do this at the expense of dismantling existing excellence, or of creating too many subcritical or substandard activities. Past trends in academic research support have resulted in a gradual but steady decentralization of academic research and graduate education, and the most rapid growth of support has taken place in institutions with no previous tradition of graduate education.* This trend

[11] "Profiles of Ph.D.'s in the Sciences, Summary Report on Follow-up of Doctorate Cohorts, 1935–1960," Publication No. 1293, NAS–NRC (Washington, D.C., 1965).

* This is factually true despite some considerable folklore to the contrary. Examples of some of the relevant statistics are as follows. In FY48, 49, and 50, eleven universities accounted for 50 percent of federal research funds. By FY63, 50 percent of the research funds were going to twenty universities. Similarly sixty-five universities accounted for 90 percent of the funds in the early period, while over 100 accounted for 90 percent of the funds in the recent period. In the period 1945–1949, ten institutions accounted for 46 percent of the Ph.D. awards in the natural sciences. By 1961 the first ten institutions accounted for only 35 percent of the awards, and by 1964 this had dropped to less than 30 percent. Between the period

should be continued, and if sufficient funds are available, should be accelerated. Current emphasis on traineeships, institutional grants, and development grants should accelerate the trends of the 1950s. However, redistribution should take place by differential growth rather than by reallocation.

CRITERIA FOR SUPPORT OF ACADEMIC RESEARCH

As implied in the previous discussion, the overall standard which should be applied to judge the adequacy of the level of academic research support is related to the growth of graduate education and the production of Ph.D.'s in science and engineering. However, academic research should not be supported for exclusively educational reasons. It must be good research as judged either by standards intrinsic to a scientific discipline or by criteria of social usefulness. If this is not the case the university role in maintaining the intellectual standards of the society will be eroded. The guiding principle of national policy should be that research should be supported in conjunction with higher education whenever it is of appropriate character, and when a realistic choice is available between academic and nonacademic support. Even when, owing to the character of the

---

1940–1949 and 1960–1961 the two regions of the country showing the greatest gains in share of Ph.D. production relative to share of population were the South Central and Mountain states, while the two regions showing the greatest relative decline were the northeast and Pacific coast regions. The maximum spread between regions dropped from a factor of 14 in the early period to only 4 in the later period and to less than 3.5 by 1964. Thus all evidence points to an accelerating trend toward dispersion of research and graduate education brought about by federal support of academic research.[12]

[12] "The Impact of Federal Research Programs on Higher Education," prepared by the Office of Science and Technology for the Research and Technical Programs Subcommittee of the House Committee on Government Operations (June 17, 1965); "NSF Funds for Academic Science, Part II, Statistical Analysis," National Science Foundation (October 1965), prepared for the House Committee on Science and Astronautics.

research, it is considered inappropriate for an academic institution, an effort should be made to locate it near a university. For academic research, however, the fields in which support is provided should be biased more toward the judgment of individual scientists as to what is important rather than by *a priori* and centrally determined judgments regarding national needs. In the terminology of A. M. Weinberg, scientific merit should be given greater weight than social merit in the support of academic research. This is in contrast to nonacademic and national center research where social merit and impact on technology are of greater importance. Academic research should be biased toward national needs only indirectly through publicizing and analyzing such needs and consequent career opportunities. As far as possible the government should depend upon this "market" information to generate new ideas and proposal pressure from faculties and fellowship applications from students. Academic research should not be biased toward national needs except to the degree that really good new scientific ideas are generated to meet these needs. This policy takes maximum advantage of the individualistic and freewheeling character of academic science and is best designed to maximize its contribution to the needs of society in the long run. Where a more highly programmed effort is desirable or where the social need is so urgent that some technical effort is needed even though there are no evident promising new approaches, it should probably be centered in nonacademic institutions, with academic participation only when interest or new ideas appear spontaneously from the academic community.

Forward planning for academic research is needed primarily in the area of facilities and equipment, and in anticipation of the need for national centers, not in allocation of resources among basic research fields at the level of "little science." In general, allocations to little science should not be made *a priori* between scientific disciplines, but should be based on the promise of individual projects, programs, or research groups. This suggestion is based on the assumption, which I believe to be true, that the variation of quality between individual projects or investigators within a discipline is far greater than the variations between disciplines as a whole. This does not necessarily mean that the role of supporting agencies should be purely passive, waiting for good ideas to come in, but it does imply that the main direction of academic research should come up from the level of the working scientist rather than down from above through central decisions of program officers, committees of scientists, or politicians. The role of supporting agencies, especially the mission-oriented ones, should be to make the academic community aware of new problems and needs and to foster communication among disciplines and problem areas through special meetings and other mechanisms, but not to "direct" research or "force feed" certain areas and starve others on the basis of *a priori* judgments of relative importance.

Criteria which should be considered in judging individual research proposals may be summarized as follows:

1. What is the promise of significant scientific results from the proposed project? The evaluation of such promise implicitly involves the past accomplishment of the investigator, and the judgment of his competence and originality by peers, either nationally or locally. The term "significant" may refer either to scientific significance or to potential applicability, but it implies some degree of fundamentality and generalizability.
2. How novel or unique is the work proposed? To what degree does it break new ground? To what extent does it exploit a new technique or unexplored research methodology? Does it provide a meaningful test of current theory and understanding in its field?
3. To what extent are the probable results of the proposed work likely to influence other work either in the same field or in related or even distant fields?

4. What is the probable educational value of the research, based on the quality and number of students or other trainees in relation to the cost of the project, the record of success of past students of the investigator, and the general academic environment in which the work is to be done?
5. What is the potential relevance of the work to possible future applications, especially in relation to existing national goals? This question is of particular relevance in connection with engineering research and with applied research in health, agriculture, or environmental pollution or other national goals.

QUANTITATIVE GROWTH REQUIREMENTS

A 15 percent annual increase in the total support, explicit and hidden, for academic science appears to be the minimum necessary to keep up with the present increase in graduate students and faculty and at the same time not lower present standards of quality or change drastically the present distribution among fields. Obviously, there are wide variations in the cost of producing a Ph.D. in different fields, and the growth of student population could in principle be accommodated by shifting present research support from the more expensive fields such as nuclear physics and astronomy into the cheaper fields such as pure mathematics or chemistry. Possibly marginal adjustments of this sort should be considered if the present stagnation of academic research budgets continues, but it would appear there are many other nonacademic federal technical activities which ought to be carefully scrutinized before any such drastic step is taken for academic science. Furthermore, the 15 percent requirement is really predicated on expanding research activity in approximately the present institutional pattern. To the extent that wider dispersion accompanies expansion, additional funds will be required to accelerate the development of research capabilities faster in the weaker institutions. So far there are no convincing estimates of what this differential is. A rough rule of thumb seems to be that on the average it costs about twice as much in the first year to start a research project in little science in a new location than it does to support a project where there is already work going on in the same or a closely related field. At present the institutional development grants, such as the NSF science development program, are really designed to provide this differential. The extra federal cost may be partly offset by the unusually high leverage on nonfederal sources of funds apparently possessed by competitive development grants.

The introduction of new technology into basic research is another factor of uncertainty. Experience in biomedical research suggests that a research field can absorb very large annual increments in funding without sacrificing scientific quality if sophisticated and reliable instrumentation is available commercially. The application of computers to almost all phases of research, including the social sciences, represents a large element of uncertainty in projecting future academic research requirements. Computer usage in universities has been increasing at an annual rate of about 40 to 45 percent per year since 1962, and the imminent introduction of time sharing and other interactive modes of computer usage is likely to accelerate both usage and costs.[13] Hitherto manufacturers' discounts and subsidy from university funds have kept down explicit charges to research. However, as computer charges rapidly become an increasing fraction of both university and research budgets, it is very unlikely that the costs can be covered by such "hidden" sources. Manufacturers' discounts have largely disappeared, and "normal" university budgets are no longer able to absorb as

[13] "Digital Computers in Universities and Colleges," a Report of the Committee on Uses of Computers, National Academy of Sciences–National Research Council (1966).

large a fraction of the costs, simply because they represent a higher proportion of the total university budget. Computer techniques so alter the character of research and the scope of the problems that can be successfully attacked in many fields that competitive pressures are likely to spread demand for new capability, including capability for teaching as well as research uses, much faster than has been the case with other types of more specialized instrumentation.

At current levels of support a 15 percent annual expansion implies about $200 million in increased support annually, including not only research support, but also support for fellowships, traineeships, training grants, and various forms of institutional or departmental support.* This "demographic" requirement, however, should not be interpreted as implying equal division among investigators, or even among institutions. It is only a statistical average based on past experience. Using such an aggregate as a standard against which to measure the adequacy of support of individual projects is not the same thing as distributing the increment according to some formula based on numbers of students, faculty, or some similar mechanical criterion. The maintenance of selectivity among projects and investigators is essential to the maintenance of quality standards in the system as a whole. Furthermore, different projects have legitimately differing costs, even within a field, let alone among fields. For example, the Chemistry Panel of the National

* Academic research has expanded from about 10 percent to 13 percent of higher education expenditures in the last ten years, and this is likely to continue as graduate education expands relative to undergraduates; Ph.D. awards in science and engineering have expanded from 4.4 percent to 8.0 percent of B.S. awards during the same period. Total costs of higher education from all sources are now expanding at an annual rate of about 12 percent. Considering the more rapid rate of expansion of Ph.D. than B.S. output, the suggested figure of 15 percent for academic research seems, if anything, rather conservative.[14]

[14] "Scientific and Technical Manpower Resources, Summary Information on Employment, Characteristics, Supply and Training," NSF 64–28 (Washington, D.C., November 1964).

Academy of Sciences made a study of research grant support in chemistry departments according to magnitude of annual support per investigator. This study showed that nearly half of the potential investigators received less than half the *average* annual grant support of $20,000 per investigator, while 27 percent of the potential investigators received no federal grant support at all. This is despite the fact that another study by the Chemistry Panel demonstrated that "the federal support per graduate student per year in chemistry is fairly uniform throughout the United States, averaging about $6000, almost without regard to the size or the scientific tradition of the university in which the student is enrolled." [15] This uniformity is probably unique to chemistry because it has the character of "little science." A study of the size of basic research support grants in NSF in 1963 showed that roughly 12 percent of the funds were expended in 1.4 percent of the grants, while 60 percent of the money was expended in about 25 percent of the grants.[16] As another illustration, Table I shows an estimate of

TABLE I FEDERAL EXPENDITURES IN UNIVERSITIES PROPER PER PH.D. PRODUCED, FY63

Astrophysics, Solar-System Physics, Cosmic Rays	500 K.*
Atomic and Molecular Physics	93 K.
Elementary-Particle Physics	910 K.
Nuclear Physics	234 K.
Plasma Physics	230 K.
Solid-State and Condensed Matter	160 K.
Chemistry	39 K.

* K = thousands of dollars.

the average cost in federal research support of producing a Ph.D. in the various subfields of physics, and also for chemistry as a whole.

[15] Reference 2, p. 168.
[16] Unpublished analysis of list of NSF grants made in FY63.

DISCUSSION

It is possible, of course, to question many of the assumptions on which the above projections are based. For example, one may question whether a research degree is really essential for undergraduate college teaching and whether in fact it imparts an inappropriate set of values to the potential undergraduate teacher, especially the teachers in four-year institutions. The same question might be raised with respect to potential industrial scientists and engineers. Indeed, the Physics Survey Panel explicitly raised the question of the desirability of an intermediate graduate degree, more appropriate for physicists who would not be engaged primarily in basic research. It is certainly true that graduate students who fail to complete their research often become valuable college teachers and are nearly as successful in industry as Ph.D. holders. However, many of them have already had research experience. Certainly, if the Ph.D. were made a much more elite degree, the necessary investment in academic research purely for educational purposes would be reduced. It must not be overlooked, however, that such a policy would greatly reduce the attractiveness of university faculty positions, and it is highly doubtful whether universities could retain intellectual leadership in society under these conditions. Few people would wish to take the risk of such a drastic departure from the past pattern in higher education.

It might also be argued that not all federally supported academic research is really essential to the training of graduate students, or that the large numbers of postdoctoral research workers in university science departments are luxuries which the budget for academic research can no longer afford. This is a complex issue with no simple answers. One of the obstacles to a simple answer is the fact that the value of graduate training depends on the total environment of the university, on the research atmosphere which is created by the presence of individuals at various levels of training. Even professors who do not supervise graduate students directly may make an important contribution to the educational environment through their influence on their colleagues who do teach. Postdoctoral students for the most part stay for only a short time and then move on into faculty positions or into industry or government. There is little evidence for the existence of a large career research population in universities divorced from the rest of the intellectual function, although clearly this is a matter that requires further attention from the universities themselves. It is clear that the competitive nature of basic research forces all researchers to demand a similar standard of living with respect to both postdoctoral associates and sophisticated instrumentation. On the other hand, it is not only other universities but also the nonacademic institutions that set the competitive standard. On the face of it, it appears rather unwise to stint on academic research when it accounts for only 7 percent of national expenditures for R and D and even only 32 percent of federal expenditures for basic research, especially if intellectual leadership in learning, as well as merely production of Ph.D.'s, is taken seriously as one of the primary social functions of universities.

If academic research budgets continue to level off, grave questions of policy are posed. The vigor of a field of science seems to depend on a continuing injection of new investigators with fresh ideas, and on sufficient funds to exploit new ideas and replace outmoded equipment. The case of controlled thermonuclear research provides an interesting model for what happens in a field in which support has stagnated. In a few short years, the United States has lost the commanding leadership which it enjoyed in high-temperature plasma research, and a principal reason was the lack of funds for new investigators and new experiments. The bad effects of this stagnation might have been less if it had been foreseen and planned for. Also there were other factors, such as a tradition of isolation from the general physics community and the absence of a base in the universities, which contributed to the stagnation of this field. The fact remains, however, that there are few, if any, examples of fields which have managed to retain their vigor and intellectual vitality in the face of level or declining support. In the absence of new funding, it will be necessary to invent new mechanisms of funding which will permit greater concentration and specialization of effort. Only by such concentration could strong growing points be maintained in the absence of new funds. A situation in which the same funds were spread more and more thinly over a growing number of investigators, institutions, and students would be only a prescription for the slow strangulation of U.S. science. Unfortunately, the necessities of maintaining quality and progress under conditions of level funding and rising costs involve policies which would be politically very unpopular. They would run directly counter to present pressures to spread federal support more widely and uniformly. In addition to fostering greater concentration of effort in a few places, it would probably be necessary to find ways of deliberately gambling on new ideas, new approaches, and new investigators at the expense of approaches or investigators whose previous record of accomplishment was high but predictable. To the extent that slowing the growth of academic research slows the rate of evolution of science, it will be necessary to try to compensate for this by gambling on mutations. The problem of maintaining the quality and momentum of science in the face of slackening growth has little precedent, and probably represents the most serious crisis faced by the U.S. scientific community in many years.

EIGHT □ Scientific Concepts and Cultural Change

The following chapter was prepared for a conference on science and culture organized by the American Academy of Arts and Sciences in the spring of 1964, and aided by a grant from the National Science Foundation. The paper was then published in revised form in the winter 1965 issue of Daedalus, *and was reprinted in a volume entitled* Science and Culture, *under the editorship of Professor Gerald Holton. I am indebted to the editor of* Daedalus, *Dr. Stephen Graubard, and to the Houghton Mifflin Company for permission to reprint this article.*

There are many difficulties of communication between the subgroups within our culture—for example, between the natural sciences, social sciences, and humanities. But there are also ways in which they are becoming increasingly united, and most of this essay will be an effort to trace a few common themes and viewpoints derived from science which I see as increasingly pervading our culture as a whole.

Perhaps one of the most important is the common allegiance of scholarship to the ideal of objective research, to the possibility of arriving by successive approximations at an objective description of reality. Whether it be concerned with the structure of a distant galaxy or the sources of the art of a nineteenth-century poet, there exists a common respect for evidence and a willingness to follow evidence wherever it leads regardless of the preconceptions or desires of the scholar. This is, of course, only an ideal; but failure to conform to this ideal, if detected, damns a scholar whether he be a scientist or a humanist. In a sense the whole apparatus of academic scholarship is an attempt to bring scientific method into the pursuit of knowledge through progressive refinements in the uncovering and use of evidence.

A characteristic of scholarship, as of science, is that it prefers to tackle well-defined, finite problems that appear to be soluble with the methods and evidence available. This often means eschewing the more fundamental, the more "metaphysical" issues, in the belief that the cumulative result of solving many smaller and more manageable problems will ultimately throw more light on the larger issues than would a frontal attack. One of the paradoxes of modern science has been that the greater its success in a pragmatic sense, the more modest its aims have tended to become in an intellectual sense. The goals and claims of modern quantum theory are far more modest than those of Laplace, who believed that he could predict the entire course of the universe, in principle, given its initial conditions. The aim of science has changed from the "explanation" of reality to the "description" of reality—description with the greatest logical and aesthetic economy. The claims to universality of nineteenth-century physics have been replaced by a greater awareness of what still remains to be discovered about the world, even "in principle." The day of global theories of the social structure or of individual psychology seems to have passed. Experience has taught us that real insight has often been achieved only after we were prepared to renounce our claim that our theories were universal. The whole trend of modern scholarship has been toward greater conservatism in deciding what can be legitimately inferred from given evidence; we are more hesitant to extrapolate beyond the immediate circumstances to which the evidence applies. We are quicker to recognize the possibility of unrevealed complexities or unidentified variables and parameters. Even in artistic criticism we tend to recognize greater diversity in the influences playing on an artist, greater ambiguity in his motives or artistic intentions. Art, scholarship, and science are united in looking further behind the face of common-sense reality, in finding subtleties and nuances. It is, of course, this search for subtlety which has made communication between disciplines more difficult, because to the casual observer each discipline appears to be working in an area beyond common sense.

The admission of finite aims in scholarship has been connected with an increasingly sophisticated view of the scope and limitations of evidence in all fields. But the emphasis on finite and limited aims in scholarly inquiry has also been paralleled by the extension of scientific and scholarly attitudes to practical affairs. One sees a close analogy between the preoccupation of science with manageable problems and the decline of ideology and growth of professional expertise in politics and business. One of the most striking developments of the postwar world has been the increasing irrelevance of political ideology, even in the Soviet Union, to actual political decisions. One sees the influence of the new mood in the increasing bureaucratization and professionalization of government and industry and in the growth of "scientific" approaches to management and administration. The day of the intuitive entrepreneur or the charismatic statesman seems to be waning. In a recent volume of *Daedalus* on "A New Europe?" the recurring theme is the increasing relegation of questions which used to be matters of political debate to professional cadres of technicians and experts which function almost independently of the democratic political process. In most of the Western world the first instinct of statesmanship is to turn intransigent problems over to "experts" or to "study groups." There appears to be an almost naïve faith that if big problems can be broken down sufficiently and be dealt with by experts and technicians, the big problems will tend to disappear or at least lose much of their urgency. Although the continuing discourse of experts seems wasteful, "Parkinsonian," the fact remains that it has worked surprisingly well in government, just as it has in science and scholarship. The progress which is achieved, while slower, seems more solid, more irreversible, more capable of enlisting a wide consensus. Much of the history of social progress in the twentieth century can be described in terms of the transfer of wider and wider areas of public policy from politics to expertise. I do not believe it is too fanciful to draw a parallel between this and the scientific spirit of tackling soluble problems.

The trend toward the acceptance of expertise has been especially striking in Europe, where both ideology and the apolitical professional bureaucracy have been stronger than in the United States. But even in this country there has been increasing public acceptance of expert analysis and guidance in such areas of government as fiscal policy and economic growth. In the realm of affairs, as in the realm of knowledge, the search for global solutions or global generalizations has been replaced by the search for manageable apolitical reformulations of problems. The general has been replaced by the specific. Concern with the theoretical goals and principles of action has been replaced by attempts at objectively predicting and analyzing the specific consequences of specific alternative actions or policies. Often the problems of political choice have become buried in debates among experts over highly technical alternatives.

It remains to be seen to what degree this new reign of the bureaucrat and the expert reflects the influence of science and scientific modes of thinking and to what degree it represents a temporary cyclic phenomenon resulting from unprecedented economic growth and the absence of major social crises. However, the modes of thought which are characteristic of science have penetrated much deeper into scholarship and practical affairs than the hand-wringing of some scientists would tend to suggest, and the general adoption of these modes of thought does not appear to have relegated genuine human values to the scrap heap to the degree which some of the humanists would have us believe. Indeed it has brought us closer to a realization of many of the human values which we regard as desirable.

On the other hand, it must be recognized that some of this reliance on expertise has moved us in directions in which we would not have gone had we been more aware of the unspoken and unrecognized assumptions underlying some of our "technical" solutions. For example, economic growth and technology have come to be accepted as valuable in themselves. The assembly line has brought more and more goods to more and more people, but it has also introduced monotony into work and a sometimes depressing standardization into our products. The technology of production tends to accept as its goals values which technology alone is well adapted to achieving without balanced consideration of other, equally important goals.[1] The very definition of gross national product connotes measurement of economic progress in purely quantitative terms without reference to changes in the quality of the social and physical environment or improvement and deterioration in the quality and variety of the products available. The inclination to tackle the soluble problems first often extrapolates to the view that the more intractable problems are less important.

In the preceding paragraphs I have argued that both scholarship and practical affairs have increasingly adopted the spirit and mode of thought of the natural sciences. An interesting question is to what extent the actual concepts and ideas of science have entered into other disciplines and into our culture generally. There are, of course, some very obvious ways in which this has occurred. Scarcely any other scientific theory, for example, has influenced literature and art so much as Freud's psychoanalytical theory. Though some of Freud's ideas might be said to contain dogmatic elements which are essentially nonscientific or even antiscientific in spirit, nevertheless, psychoanalysis is based on largely empirical observation and professes to test itself against objective evidence. It is clearly a scientific theory which, though extensively elaborated and modified, is still basically valid in its description of the irrational and subconscious elements in human motivation and behavior. It has completely altered our view of human nature, and this changed viewpoint is reflected almost universally, though in varying degrees, in modern literature and art, as well as in the interpretation of history and political behavior. The orderly Lockean

[1] Lewis Mumford, "Authoritarian and Democratic Technics," *Technology and Culture*, 5 (1964), 1.

world embodied in the American Constitution, in which each man acts rationally in his own self-interest, can no longer be accepted in quite the undiluted way that the Founding Fathers believed in it. There is ample evidence of neurotic and irrational behavior on the part of whole communities and social systems, often in opposition to their own self-interest. Even organized religion has largely accepted and adapted many of the principles of psychoanalysis, while rejecting some of the world views which have been extrapolated from it.

A more problematic example is the parallel between the increasingly abstract and insubstantial picture of the physical universe which modern physics has given us and the popularity of abstract and nonrepresentational forms of art and poetry. In each case the representation of reality is increasingly removed from the picture which is immediately presented to us by our senses. As the appreciation of modern physics requires more and more prior education, so the appreciation of modern art and music requires a more educated — some would say a more thoroughly conditioned — aesthetic taste. In physics the sharp distinction which used to be made between the object and its relations to other objects has been replaced by the idea that the object (or elementary particle) is nothing but the nexus of the various relations in which it participates. In physics, as in art and literature, form has tended to achieve a status higher than substance.

It is difficult to tell how much psychological reality there is to this parallel. It is not sufficient to reply that a physical picture is still a definite model which can be related by a series of clear and logical steps to the world which we see and that no such close correspondence exists between abstract art and the sensible world. For physical models depend to a larger degree on taste than is generally appreciated. While correspondence with the real world exists, this probably is not sufficient by itself to constitute a unique determinant of a model. Yet the successful model is one that has evolved through so many small steps that it would take a bold imagination indeed to construct another one which would fit the same accumulation of interconnected facts or observations. What is regarded as acceptable evidence for a model of reality, even in physics, is strongly dependent on the scientific environment of the time. Evidence which favors theories already generally accepted is much less critically scrutinized than evidence that appears to run counter to them. One always makes every effort to fit new evidence to existing concepts before accepting radical modifications; if a theory is well established the contradictory evidence is usually questioned long before the theory, and usually rightly so. Established theories depend on many more bits of accumulated evidence than is often appreciated, even by the scientist himself. Once a principle becomes generally accepted the scientific community generally forgets much of the detailed evidence that led to it, and it takes a real jolt to lead people to reconsider the evidence. In fact, scientific theories are seldom fully displaced; rather they are fitted into the framework of a more comprehensive theory, as Newtonian mechanics was fitted into the formulations of relativity and quantum mechanics. This in itself suggests that there are many theories or models which will fit given facts.

All of this points to the fact that a scientific theory is the product of a long evolutionary process which is not strictly logical or even retraceable. The mode of presentation of science, especially to the nonscientist, usually suppresses or conceals the process by which the results were originally arrived at, just as the artist does not reveal the elements which went into his creation. Thus it seems possible that there is some common or universal element in the modern mentality which makes quantum theory acceptable to the physicist, abstract art to the artist, metaphysical poetry to the poet, atonal music to the musician, or abstract spaces to the mathematician. The attack on these aspects of modern culture by totalitarians of both the right and the left perhaps lends further credence to these common threads. It is interesting to observe that children with previously untrained tastes have little trouble in appreciating and enjoying modern art or music and that the younger generation of physicists has no trouble in absorbing the ideas of quantum mechanics quite intuitively with none of the sense of paradox which still troubles some of the older generation. It is probable that the main elements of taste, whether it be scientific or aesthetic, are formed quite early in our experience and are strongly conditioned by the cultural climate. Science, as one of the most dynamic of contemporary intellectual trends, is undoubtedly a strong factor in creating this cultural climate, but it would be rash to ascribe causal connections. It would be interesting to know whether some psychologist, by studying current tastes in art or poetry, could predict what *kinds* of theories were likely to be acceptable in elementary-particle physics, or perhaps vice versa!

Another obvious but superficial way in which scientific ideas enter our culture is through some of the dominant "themes" of science. One such theme, for example, is evolution and natural selection, and the derived philosophical concept of progress. Today we take the idea of evolution so much for granted that we are inclined to forget that until the nineteenth century it was generally believed that the present state of society and man was the result of degeneration from some antecedent golden age or hypothetical ideal "state of nature." The Puritan Revolution in England and the French Revolution had ideologies which appealed to a hypothetical prehistoric past for their model of an ideal society. Only with Marx did revolution present itself as a forward movement into a more "advanced," previously nonexistent state of human society.

In the nineteenth century the idea of evolution and particularly the concepts of natural selection, competition between species, and the "survival of the fittest" were seized upon as an explanation of and justification for the contemporary laissez-faire capitalist society. State intervention in the competitive economic process was regarded as an almost immoral interference with the "balance of nature" in human society. In the United States and Britain the first science of sociology was built upon an interpretation of the ideas of natural selection. A whole generation of future American businessmen was educated in the ideas of men like Sumner. This sociology stressed the dangers of permitting organized society to tamper with the inexorable laws of social evolution.

In the early part of the twentieth century Darwin's ideas lost some of their original influence, but now, in the second half, they have regained much of their influence in biology and have tended to be reinforced by recent discoveries in biochemical genetics. However, it is interesting to note that a subtle change of emphasis has crept into the interpretation of natural selection. The modern evolutionary biologist tends to stress the concept of the "ecological niche" and the fact that natural selection, when looked at more carefully, leads to a kind of cooperation among species, a cooperation which results from finer and finer differentiation of function and of adaptation to the environment.[2] Indeed, biologists stress the fact that natural selection generally leads not to the complete domination of one species, but rather to a finer and finer branching of species, a sort of division of labor which tends ultimately to minimize competition. Is it too much to suggest a parallel here between the changing scientific interpretations of biological evolution and changing attitudes toward cooperative action in human societies? Is there any connection between the modern view of ecology and the progressive division of labor and specialization of function which are characteristic of modern economic organization? Certainly the analogies with biological evolution have

[2] Ernst Mayr, *Animal Species and Evolution,* Cambridge: Harvard Belknap Press (1963).

been extremely suggestive in the development of modern cultural anthropology.

Another theme which is involved here is that of dynamic equilibrium or balance, also fruitful in the study of chemical equilibrium. When dynamic equilibrium exists, a complex system can be apparently static from the macroscopic viewpoint even though rapid changes are taking place in its elementary components. All that is necessary is that the rates of changes in opposite directions balance. This is the kind of equilibrium that is envisioned as occurring in an ecological system or in a social or economic system. It would, perhaps, be wrong to suggest any causal or genetic relation between the growth of such ideas of chemical theory and their application to social or biological systems. The fact is, however, that the concepts arose at similar periods in scientific development and helped to establish a kind of climate of taste in scientific theories which undoubtedly facilitated intuitive transfer from one discipline to another. One finds the images and vocabulary of chemical equilibrium theory constantly recurring in descriptions of social and economic phenomena.

Two of the germinal ideas of twentieth-century physics have been "relativity" and "uncertainty." Philosophers generally recognize that both of these themes have had an important influence on their attitudes, but the physical scientist finds it more difficult to connect the philosophical view with its role in physics. At least the connection is not so self-evident as it is in the case of evolution or of psychoanalysis. Indeed, both relativity and uncertainty are words which have rather precise operational meanings in physics, but which have been given all sorts of wishful or anthropomorphic interpretations in philosophy. Indeed, scientific popularizers have themselves been especially guilty of this type of questionable semantic extrapolation.[3] The situation

[3] L. S. Stebbing, *Philosophy and the Physicists,* New York: Dover Publications, Inc. (1958).

has been aggravated by the tendency of physicists to use words from everyday discourse to denote very subtle and precise technical concepts. The popularizer and the layman then use the technical and the everyday term interchangeably to draw conclusions bearing little relation to the original concept.

Let us consider relativity first. The basic idea of relativity is that all the laws of mechanics and electromagnetism are the same, independent of the state of uniform motion in which the observer happens to be moving. Relativity is "relative" in the sense that there is no "absolute" motion, no fixed reference point in the universe that has greater claim to validity than any other. On the other hand, the elimination of absolute motion is achieved only at the price of introducing an absolute velocity which is the same in all reference systems, namely, the velocity of light. Thus it may be legitimately questioned whether "relativity" or "absolutism" is the correct name for the theory. Nevertheless, the first terminology was the one that caught the popular and speculative imagination and provided the basis of a revolution in viewpoint which affected many areas of knowledge. Not long after relativity was absorbed into physics, the anthropologists were stressing the extraordinary diversity of human customs and ethical norms and were arguing that moral standards had to be viewed not in an absolute sense but relative to the particular culture in which they were found. The judgments of history became less moralistic; the actions of individuals tended to be viewed in the context of the ethical norms of their time. The realistic novel or drama in which human behavior was depicted without moral judgment became fashionable. Yet if these things have little to do with "relativity" in the sense that Einstein intended, the very fact that the word caught fire so easily suggests there does exist a kind of common taste in such matters and that this taste forms part of the intellectual climate of the time.

The other key idea of physics is "uncertainty," as embodied

in the Heisenberg Uncertainty Principle. The philosophical interpretation of this principle has been the subject of interminable debate by both scientists and laymen. On one extreme, people have viewed the uncertainty principle as repealing the laws of causality and reintroducing "free will" into the physical as well as the mental universe. Most working physicists tend to take a somewhat more pedestrian view of the principle. They interpret it as being the result of an attempt to describe the state of the universe in terms of an inappropriate and outmoded concept, namely that of the point mass or "particle," a concept derived by analogy with macroscopic — that is common-sense — physics. Nevertheless, regardless of the exact interpretation, the uncertainty principle does imply that the idealized classical determinism of Laplace is impossible. The laws of quantum theory are deterministic or "causal" in the sense that the state of the universe at any time is determined by its "state" at some previous time. The lack of determinism in the Laplacian sense comes from the impossibility of specifying the "state" at any time in terms of any set of operations which will not themselves change its state and thus spoil the assumptions. What is wrong in the old determinism is the idea that the universe can be uniquely and unequivocally distinguished from the observing system, which is a part of it. In this sense the uncertainty principle can be seen as merely a further extension of the concept of relativity.[4] Interpreted in this light, we find the same idea cropping up in many fields of knowledge. The social scientist is increasingly conscious that the measurements that he can make on any social system affect the future behavior of the system. A good example is public opinion polls, which, if made public, affect the attitude of the public on the very matters the polls are

[4] P. W. Bridgman, *The Logic of Modern Physics*; first edition, New York: Macmillan (1927). Although written as a philosophical interpretation of relativity before the discovery of quantum mechanics in its modern form, this work is extraordinarily prescient with respect to the philosophical ideas underlying quantum theory.

supposed to measure "objectively." Another example is educational tests, which not only measure human ability, but tend to change the cultural and educational norms which are accepted and sought. This aspect of the uncertainty principle in the social sciences is, in quantitative terms, a matter of some debate, but it is an important factor in social measurement, which has to be dealt with just as in physics. In many social situations the mere fact that the subjects know they are being observed or tested affects their behavior in ways which are difficult to discount in advance. Even in a subject like history a sort of analog of the uncertainty principle is found. It lies basically in the fact that the historian knows what happened afterwards and therefore can never really describe the "initial conditions" of his system in a way which is independent of his own perspective. In seeking to discern the underlying causes of events he inevitably tends to stress those factors which demonstrably influenced events in the way they actually came out, minimizing factors or tendencies which did not develop even though the relative strengths of the two tendencies may have been very evenly balanced at that time. The modern historian, of course, tends to be very aware of this uncertainty principle and to allow for it as much as possible. Again, while there is probably little intellectual connection between these various attitudes in the different disciplines, there is a general intellectual climate which stresses the interaction between the observer and the system being observed, whether it be in history, physics, or politics.

There are a number of themes in science having a somewhat more direct and traceable intellectual connection between different disciplines. Here I should like to mention three, namely, energy, feedback, and information. Each of these is a highly technical concept in physics or engineering; however, each also has broad and increasing ramifications in other disciplines. Of these, the oldest and most loosely used is probably energy. This concept is closely associated with that of "transformation."

That is, the reason energy is a useful concept is that it has many different forms or manifestations which may be transformed into each other. In physics it is probably the most general and unifying concept we have. All physical entities or phenomena, including "matter" or "mass," are forms or manifestations of energy. Though it may be transformed, its quantity is "invariant," and this is what makes it important. The concept of energy has, of course, been important in biology almost as long as in physics. Living matter functions by transforming energy, and much of the early science of physiology was concerned with studying the transformations of energy in living systems. But the term "energy" has also found its way into many other fields of knowledge, where it is used often more metaphorically than with precise significance. Nevertheless, even in its metaphorical use it tends to partake of some of the characteristic properties of physical energy; namely, it is subject to transformation into different forms, and in the process of transformation the total energy is in some sense preserved. One speaks of psychic energies, historical energies, social energies. In these senses energy is not really measurable, nor is it directly related to physical energy. Nevertheless, like physical energy it can be released in the form of enormous physical, mental, or social activity; and, when it is, we tend to think of it as somehow "potential" in the pre-existing situation. The term "tension" denotes a state of high potential energy, like a coiled spring; and a high state of tension, whether social or psychological, is usually followed by a "release" or conversion into kinetic energy or activity of some variety. Thus the language of energy derived from physics has proved a very useful metaphor in dealing with all sorts of social and psychological phenomena. Here the intellectual connection is more clear than in the case of relativity or uncertainty, but it is more metaphorical than logical.

The concept of feedback is one of the most fundamental ideas of modern engineering. It underlies the whole technology of automatic control and automation. The original concept was quite restricted in application. It arose in connection with the design of electronic amplifiers in which a part of the output was fed back into the input in order to control the faithfulness with which the amplifier would reproduce in the output the form of the input signal.[5] An amplifier with what is called negative feedback reproduces the time behavior of the input more faithfully than the same amplifier without feedback, the more so the greater the feedback.

The concepts and methods of analysis originally developed for amplifiers were rapidly applied to control systems, where they had a far more fundamental influence. In recent years the feedback concept has been extended still further to embrace the idea of "information feedback," which is important in biological and social phenomena as well as in the engineering of physical systems. The idea has been stated by Forrester[6] in the following way:

"An information feedback system exists whenever the environment leads to a decision that results in action which affects the environment and thereby influences future decisions."

At first this may seem unrelated to amplifiers and control systems, but if we identify "environment" with "input" and "decision" with "output" we can readily see how the more general definition includes amplifiers and control systems as a special case. In the case of the amplifier the decision is completely and uniquely determined by the environment, but the concept of information feedback applies equally well when the decision is a discrete rather than a continuous function and when it is related to the environment only in a probabilistic sense.

[5] H. S. Black, U.S. Patent No. 2,102,671 (1934); cf. also, H. W. Bode, *Network Analysis and Feedback Amplifier Design*, New York: van Nostrand (1945).

[6] J. W. Forrester, *Industrial Dynamics*, Cambridge and New York: M.I.T. Press and John Wiley (1961).

In this more general use of the words environment and decision we can see many examples of the information feedback concept in biology and the social sciences. For example, the process of natural selection in evolution is itself a type of feedback. The selection process — the particular population which survives in each generation — is the decision, and this is fed back into the genetic constitution of the next generation; in this way the characteristics of the population adjust to the environment over successive generations.

The muscular activities of animals also illustrate information feedback. In this case, the environment, which must be considered as including both the external environment and the relation of the body to it, influences the decision through perception. The work of the muscles is the analog of the amplifier or controller, and the perception of the organism provides the feedback loop. The process of learning may be readily regarded as an information feedback system. Indeed the theory behind the teaching machine is essentially designed to establish a tighter feedback, through the process of "reinforcement," which helps the student to decide whether he has learned correctly. Much of the concern with the techniques of teaching is related to improvement of the feedback loop in the learning process.

The consideration of such processes as learning or cultural evolution as feedback systems would be merely a convenient metaphor, like energy, were it not for the fact that information feedback systems have certain general properties which tend to be independent of their particular embodiment. The two most important properties are those of stability and response. There exists a whole theory of the stability of feedback systems, which depends on the amplification or "gain" of the system and the time delays which occur throughout the whole decision-environment-decision loop. High gain and large time lags tend to produce instability which will cause the system as a whole to "hunt," that is, the state of the system oscillates in a more or less uncontrolled way about the position of adjustment to the environment. The term "response" relates to the closeness and rapidity with which the system in question will adjust to a changing environment; this is analogous to the closeness with which the time behavior of the output of a feedback amplifier will reproduce the time behavior of the input.

The mathematics of the stability of linear amplifiers and control systems — that is, physical systems in which the output or "decision" is directly proportional to input or "environment" — is highly elaborated and well understood. Real feedback systems, however, are often nonlinear, probabilistic in nature, and discrete rather than continuous. The mathematics for dealing with such systems is not very well developed, and for this reason it has not, until recently, proved very profitable to look at biological or social systems from the standpoint of information feedback. However, the advent of the high-speed digital computer has speeded up the processes of ordinary arithmetical calculation by a factor of more than a million and has brought much more complicated and pathological (from the mathematical standpoint) systems within the purview of calculation. The usefulness of the computer lies in the fact that the behavior of feedback systems depends on only certain of their abstract properties; these properties, in turn, can be readily modeled or "simulated" on a computer. Thus we are enabled to study the dynamics of the model in great detail and, if necessary, at a speed much greater than that of the real life situation.

It is now being recognized that many types of unstable behavior that occur in biological and social systems are, in fact, examples of unstable feedback systems, the instability usually arising from unacceptable time lags in the transmission of information through the system. A case which is by now fairly well documented is that of inventory policy in a business.[7] In times of high demand a business may tend to build up inventory

[7] *Ibid.*

in anticipation of future demand, and this further increases demand; but there is a lag between orders and production as well as between the measurement of demand and the decision to increase inventory. This can have the effect of introducing a highly fluctuating factory output in a situation in which the external demand is actually rather steady. Forrester[8] has given an analysis which suggests strongly that exactly this model may account for the notorious instability of production and employment in the textile industry.

It seems highly likely that the business cycle in the economy as a whole represents a form of feedback instability to which many individual elements of decision-making contribute through their time lags. In fact all forms of social decision-making tend to contain an inherent time lag arising from the fact that anticipations of the future are simply linear extrapolations of past trends. Thus one can even discern a similar tendency in history for political and social attitudes towards public issues to be those appropriate to the experience of the recent or distant past rather than to the actual situation which is faced. For example, the philosophies of laissez-faire economics were conditioned by the mercantile and preindustrial era in which the principal problem was the inhibiting effects of state interference in the economy. Or, to take a more recent example, early postwar American economic policy was based on the fear of a major depression similar to what followed the first war, while much of present public thinking is based on the fear of inflation of the type which followed World War II. Such lags in social attitudes probably contribute to many of the cyclic phenomena which are often attributed to history. Of course, the examples given above are somewhat crude oversimplifications, but the basic idea is one which may have quantitative as well as suggestive or metaphorical value.

Another possible example is the cycle in moral attitudes. Atti-

[8] *Ibid.*

tudes toward moral values, because of the long time they take to diffuse throughout society, tend to lag behind the actual social conditions for which they were most appropriate. Thus, for example, Victorian attitudes toward sex arose partly as a reaction to the extreme laxity which existed in previous times, and conversely modern liberal attitudes toward sex are to some extent a response to the social and psychological effects of Victorian repression. Such attitudes tend to go in cycles because their inherent time lag produces unstable feedback in the social system. Such lags are especially important in the dynamic or "high gain" cultures characteristic of the West.

The problem of stability in feedback theory is relevant to situations in which the environment without considering feedback is more or less constant. When an unstable feedback situation exists, the system "hunts" about the stable situation of adjustment to the environment. The other important concept, however, is that of the response of the system to environmental changes imposed from without, or, in amplifier terminology, the faithfulness and speed with which the output follows the input. This introduces the idea of "optimization" in control systems. An optimized system is one which responds to its environment in the best way as defined by some quantitative criterion. Of course, the optimal configuration of the control system will be dependent on the properties of the environment to which it is expected to respond or adapt. We can imagine an environment which is subject to short-term and long-term changes and a feedback system which is optimized for the short-term changes occurring during a certain period. If the nature of the short-term changes also varies slowly in time, then the feedback system will not remain optimum. We could then imagine a feedback system whose properties change with time in such a way as to keep the response optimal as the short-term changes in the environment occur. The continuing optimization can itself be described as a form of information feedback. For example, we can imagine a man learning a game requiring great physical skill. When he has learned it, his muscular and nervous system may be thought of as a feedback system which has been optimized for that particular game. If he then engages in a new game, his muscles and nerves will have to be optimized all over again for the new game, and the process of learning is itself a form of information feedback. In this way we arrive at the concept of a whole hierarchy of feedback systems — of systems within systems, each operating on a different time scale and each higher system in the hierarchy constituting the learning process for the next lower system in the hierarchy. In the technical literature these hierarchies are referred to as adaptive control. Adaptive control systems appear to "learn" by experience and thus come one step closer to simulating the behavior of living systems. In fact we may imagine that biological and social systems are information feedback systems with many more superimposed hierarchies than we are accustomed to dealing with in physical control systems.

The other key idea from engineering which has had an important impact on social and biological theory is that of information, and the closely associated concept of noise. The idea that information was a concept that could be defined in precise mathematical terms was recognized by Leo Szilard in 1929.[9] Szilard was the first to point out the connection between the quantity of information we have about the physical world and the physical concept of the entropy of a system. However, Szilard's ideas lay fallow until twenty years later when they were rediscovered by Shannon[10] and precisely formulated in their modern form — the form which has revolutionized modern communications. There is a very close relation between information and probability. In fact, the amount of informa-

[9] L. Szilard, Z. *Phys.* 53 (1929), 840; cf. also, L. Brillouin, *Science and Information Theory*, New York: Academic Press (1956).

[10] C. E. Shannon and W. Weaver, *The Mathematical Theory of Information*, Urbana: University of Illinois Press (1949).

tion in an image or a message is closely connected with its deviations from a purely random pattern. The concept of information is basic to the quantitative study of language and has provided one of the cornerstones of a new science known as mathematical linguistics. It is also generally recognized that the transmission of genetic properties from generation to generation is essentially a communication of information. This has led to the idea of the "genetic code" which contains all the information necessary to reproduce the individual. So far the attention of biologists has mostly been focused on the elementary code, that is, on the relationship between the structure of the DNA molecule and the genetic information which it carries. The possibility of precise definition of the quantity of information in a system, however, opens up the possibility of considering evolution from the standpoint of a system of information transmission,[11] a type of study which is still in its infancy. A remarkable consequence of information concepts is the realization that the information embodied in the biological constitution of the human race is essentially contained in the total quantity of DNA in the human germ cells — at most a few grams in the whole world. One suspects that it may be possible to apply information concepts similarly to the study of cultural evolution and to the transmission of culture from generation to generation.

One cannot talk about information without considering noise, which is the random background on which all information must ultimately be recognized. By its very nature noise is the absence of information. When an attempt is made to transmit a definite piece of information in the presence of noise, the noise destroys a definite amount of the information in the transmission process. No transmission system is completely faithful. Noise is, in the first instance, a physical concept; but, as in the case of information and feedback, the concept may be extended in a

[11] W. Bossert, private communication.

somewhat vague way to social and biological systems. For example, in evolutionary theory the "noise" is the random variations in the genetic constitution produced by cosmic radiation and other external influences on the genetic material. In the transmission of cultural information, the "information" communicated by a piece of literature or a work of art depends not only upon the intrinsic information content of the work but also on the experience and education of the recipient. Unless the artist and the recipient have had the same experience, the communication is always less than faithful.

In the foregoing I have tried to suggest how a number of important themes from the physical and biological sciences have found their way into our general culture, or have the potential for doing so. In the case of the concepts of feedback and information, the ideas appear to have an essentially quantitative and operational significance for social and cultural dynamics, although their application is still in its infancy. The most frequent case is that in which a scientific concept has served as a metaphor for the description of social and political behavior. This has occurred, for example, in the case of the concepts of relativity, uncertainty, and energy. In other cases, such as evolution and psychoanalysis, the concept has entered even more deeply into our cultural attitudes.

NINE □ Engineering, Science, and Education

During 1961 and 1962 the Engineers Joint Council appointed a committee under the chairmanship of Dr. J. H. Hollomon to study ways in which engineering research might contribute toward solving some of the urgent problems facing our society within the next ten years. I was assigned the task, as a subcommittee of one, to deal with the relationships between engineering and the processes of education. The present chapter is a minor revision of the subcommittee report that was issued in April 1962. It deals with the coming requirements for education, with some of the ways in which engineering devices may contribute toward making learning more effective, and with the application of concepts derived from engineering theory to the understanding of the learning process. The latter repeats some of the themes mentioned in the preceding chapter on "Scientific Concepts and Cultural Change," reflecting the fact that education is one of the principal instrumentalities of cultural change.

INTRODUCTION

Educational devices such as audio-visual aids and teaching machines have been widely heralded as foreshadowing a revolution in educational techniques in the next ten years. For this reason alone, although education has not been traditionally regarded as a domain of engineering or engineers, it can as a practical matter no longer be ignored in any discussion of the future role of engineering in solving the problems confronting our society.

THE CHALLENGE OF EDUCATION

Nowhere is the challenge of the next decade more apparent than in education. Experience and recent history demonstrate that the educational level of a country constitutes its single most productive and most indestructible form of national capital. This is true not only because of the specific skills which are acquired through education, but also because education greatly increases the ability of individuals and groups to adapt to a complex, rapidly changing, and urbanized society. The dramatic recovery of the Netherlands, Germany, Japan, and the Soviet Union from the destruction of World War II is in large part attributable to the much higher educational level in these countries as compared with the period after World War I. The economic strength of Israel in the midst of the economic desert of the Arab world is another startling demonstration of the importance of education. Without question the training in technical skills received by many men in the U.S. armed forces both during and after the war, combined with the spur to education given by the G.I. Bill, has played a large part in the tremendous surge of industrial growth which this country has seen almost without a break for more than fifteen years. Doubtless there are many other favorable factors in all these examples. Education is a necessary but not a sufficient condition for economic progress and political stability.

The educational challenge facing the world, and particularly the United States, has five major aspects.

1. There is a growing realization that our present educational system does not enable most individuals to realize their full potential capabilities. This realization comes at a time when the quantitative demand for education, in the next five years at the secondary school level, and in the following five at the college level, is expected to take an upswing which will place severe strains on both the teaching profession and the financial resources available for education. The rapidly growing need of our society for the intellectual professions increases the demands on the educational system at the same time that it increases the competition for the talented individuals who are needed as teachers. We must find ways of improving the effectiveness of the educational enterprise at all levels.

2. The social dislocation generated by the rapid transformation of our economy through automation and through foreign competition is creating an unprecedented problem, of which we are seeing only the first symptoms. A large part of the solution to this problem must lie in adult education which will help people to adapt more quickly to our dynamic economy as particular skills become obsolescent. This is a problem which we scarcely understand. Certainly its solution will require entirely new educational techniques adaptable to individuals of widely varying character and background, and in which self-education must play a key role.
3. If the present arms race continues into the coming decade, the increasing complexity and rate of obsolescence of military equipment will place ever-increasing demands on training. The armed services have the same problems as industry in rejuvenating obsolete skills, but made much more acute because of the extremely dynamic character of military technology and the large peacetime turnover of personnel.
4. The problem of propelling the underdeveloped countries into the self-sustaining stage in their economic development is fundamentally one of education. In the West there is insufficient realization that these countries cannot afford to wait for the traditional evolution of Western educational systems. Bold innovation is required to achieve shortcuts and to multiply effectively the influence of the few talented teachers available.
5. The content and method of education all over the world need to be greatly improved to enable men and women to better exercise the responsibilities of freedom in an increasingly complex and interdependent world political system. Among other elements this undoubtedly involves a much better understanding of and attention to the inter-

action between educational experience and the moral, cultural, and emotional background of the people being educated.

QUANTITATIVE AND QUALITATIVE DEMAND FOR EDUCATION IN THE UNITED STATES

The ten-year projections of school and college enrollment, based on studies by the Office of Education, are shown in Table I.[1,2,3] This table shows clearly that the biggest increases in demand will come at the higher levels. This situation reflects both the increasing population in the secondary-school- and

TABLE I PROJECTIONS OF PUBLIC ELEMENTARY SCHOOL, SECONDARY SCHOOL, AND COLLEGE ENROLLMENT

Year	*Elementary*	*Secondary*	*College*
1960	24.32 million	11.66 million	3.6 million
1970	28.03	16.49	6.0
% Increase	15.3%	41.2%	66.7%

college-age groups and the relative growth of the professional and technical component of the national labor force. Our increasingly urban and technological society has an insatiable demand for all the intellectual professions. From 1950 to 1959, for example, while the labor force increased only 10.8 percent, the number of professional and technical workers increased by 61.5 percent. The professional group will increase by at least 40 percent in the next ten years, while scientists and engineers

[1] "Ten Year Aims in Education, Staffing and Constructing Public Elementary and Secondary Schools 1959–1969" (Washington, D.C.: Office of Education, released January 19, 1961).
[2] "Ten Year Objectives in Education, Higher Education Staffing and Physical Facilities, 1960–61 through 1969–70" (Washington, D.C.: Office of Education, released January 19, 1961).
[3] "Investing in Scientific Progress, 1961–1970, Concepts, Goals, and Projections" (Washington, D.C.: National Science Foundation 61–27).

will tend to be an increasing fraction of the professional and technical work force as suggested by Table II. The percentage

TABLE II PROFESSIONAL AND TECHNICAL WORK FORCE; TECHNICAL WORK FORCE AS A PERCENTAGE OF TOTAL WORK FORCE; TECHNICAL DOCTORATES AS A FUNCTION OF PROFESSIONAL WORK FORCE

Year	*Total Professional*	*Technical*	*Technical/ Professional*	*Technical Ph.D./ Professional*
	millions	millions	%	%
1950	4.46	0.8	17.9	1.01
1960	7.2	1.4	19.5	1.2
1970	10.1	2.5	24.8	1.66

increases in various categories between 1950 and 1959 are also shown in Table III, which shows a trend likely to continue.

TABLE III INCREASES IN WORK FORCE IN VARIOUS CATEGORIES 1950–1959

	%
Professional and Technical	61.5
Total Labor Force	10.8
School Instructional Staff	44.2
Science and Engineering	64.2
Doctoral Scientists and Engineers	93.0

The demands for professional educational staff implied by enrollment figures are summarized in Table IV. This table suggests that the requirement can be met quantitatively in the sense that the staffing of educational institutions of all sorts will not require a greater overall growth than is demanded for all the intellectual professions; in other words, individuals will not have to be diverted into education, but the education profession will have to attract its fair share, a requirement which will necessitate a substantial improvement in the relative economic position of the teaching profession within the next few years — according to the Office of Education, a close to 50 percent increase in average salaries at all levels combined with a much bigger range of salaries according to ability. Table IV

TABLE IV DEMANDS FOR SCHOOL AND COLLEGE STAFF, 1960 TO 1970 WITH PERCENTAGE INCREASES. ALL NUMBERS IN MILLIONS

Year	A	B	C	D	E	F
1960	.102	.834	.518	.270	.135	1.724
1970	.137	.929	.741	.405	.260	2.212
Percent	34.2%	17.1%	43.0%	50%	92.5%	28.2%

Explanation of Table IV:

A — Supervisory and other professional staff.
B — Nonsupervisory instruction staff — elementary school.
C — Nonsupervisory instructional staff — secondary school.
D — College and university, including research staff.
E — Science and engineering in college and university, including basic research staff.
F = A + B + C + D = Total professional staff.

TABLE V COST OF ALL EDUCATIONAL ACTIVITIES (ANNUAL) AS A PERCENTAGE OF GROSS NATIONAL PERSONAL INCOME

	1960	*1970*
Public Elementary and Secondary	2.12%	3.60%
College and University	.95%	1.58%
TOTAL	3.07%	5.18%

also suggests that there may be a qualitative problem, in that the expansion in technical fields is considerably greater than the average, if the projections are correct. The demand, however, could be met without putting undue strain on the requirements of other professions.

The resource demands of education at all levels are summarized in Table V. The figures are given in terms of annual expenditures as a percentage of national personal income. They include both salaries and capital expenditures, and the salary figure allows for a restoration of the relative economic status of the teaching profession at all levels as projected by the Office of Education. The table shows that education will demand an increasing share of national income, but the requirement does not appear to be unreasonable, especially when one considers that the United States spends about 20 percent of its gross national product on all classes of capital investment, to which education can be legitimately compared. Education has already substantially upped its share of national income in the last five years.

The above projections allow no credit for increased "productivity" of teachers due to improved methods and educational devices. On the other hand, neither do they take into account the probable increase in proportion of scientific and technical training, which requires more facilities and a smaller teacher to student ratio. Furthermore, it fails to take into discount the fact that the social sciences are to an increasing degree becoming laboratory sciences, requiring facilities and instruction methods more nearly analogous to the natural sciences. Thus on balance the projections are probably fairly reasonable, and they suggest that there is no quantitative crisis as regards either personnel or resources provided the political process generates an adequate resource base for education.

The question of quality is much harder to document. There is a quite general, largely undocumented, belief that few people reach the achievement level of which they are capable, or which is going to be increasingly necessary to the satisfactory functioning of our economic and political system. In its January 1961 report on ten-year objectives in education, the Office of Education states (see reference 2):

> A mere moment's reflection . . . leads to two conclusions: (1) that more youth of above average ability must be induced to enter the main stream of intellectually demanding endeavor, and (2) that the educational achievement of all youth must be so raised

that the average adult will ultimately be able to meet occupational challenges far above those that the average adult of the same age would have had occasion to meet a generation ago.

Our more recent knowledge of human development strongly suggests, if it does not conclusively demonstrate, that the ceiling of individual potential is far higher than is generally approached in actuality . . . both the survival of the nation as we know it and the fulfillment of what we tend to consider its historic mission in the world demand our continued and intensified effort to establish a higher base of educational development than we have ever thought possible.

The fact that the projected demand for the next ten years lies largely at the higher levels of education does not necessarily imply that we should concentrate all our efforts at this level. In the past it has been customary to regard elementary education as primarily a problem of methodology, and to attempt to erect a superstructure of advanced education on a foundation of elementary education that remained more or less static as to content and organization. Recent experience suggests that content is equally important at the lower levels, that the whole educational process must be regarded as an integrated process of development, that the results of scholarship on the frontiers of knowledge have important implications for education even at the most elementary level. It is thus a fault of our educational system that it has accorded insufficient importance to content at the more elementary levels and possibly insufficient importance to method at the more advanced.

There are two main streams of intellectual ferment working in education. One is represented by the Physical Science Study Committee and similar efforts in mathematics, biology, and now English. This is bringing the fruits of scholarship and the attention of advanced scholars to the problems of secondary school curricula, and will hopefully be brought to bear on the elementary school. The emphasis is on content, but teaching aids and methods are exploited in an imaginative and original way. The other stream of development is represented by the teaching machine and programmed learning.[4] This is bringing to bear the results of research in the psychology of learning on the actual methodology of teaching at all levels. It is a methodology, but one in which a profound insight into subject matter plays an indispensable part, so that again individuals working at the frontiers of knowledge must participate. The most exciting aspects of both of these developments is that they have made teaching intellectually respectable, a fitting activity for the most advanced scholars and the brightest minds. This aspect is probably far more important than the particular methods and content.

ADULT EDUCATION

It is characteristic of our present economic system that the impact of unemployment has fallen mainly on the least educated and least skilled segment of our labor force. These are the individuals who lack any special skills but also lack the educational level needed to acquire new skills. Increasingly the worker who has acquired great skill primarily through long experience in a particular trade or industry finds himself unemployed as a result of a change in technology, but finds himself also with insufficient adaptability to learn a new trade. Education imparts the ability to adapt; it is the cultural counterpart of biological adaptation, and is necessary for survival in our society. Our economy is now changing so rapidly that we can no longer await the better education of the next generation to overcome the effects of educational obsolescence. We must devise ways

[4] *Teaching Machines and Programmed Learning, a Source Book*, Ed. by A. A. Lumsdaine and R. Glaser; published 1960 by National Education Association, Washington, D.C. Following articles are referred to especially:
Articles by S. L. Pressey, pp. 32–52; B. F. Skinner, "The Science of Learning and the Art of Teaching," pp. 99–114 (1954); Simon Ramo, "A New Technique of Education," pp. 367–382 (1957); A. A. Lumsdaine, "Concluding Remarks," pp. 563–573 (1960).

to help the adult, and ways which are adapted to his needs and limitations. We suspect that this re-education will have to be done largely within industry, and a great deal of it will be the responsibility of engineers or managers with an engineering background. Many industries are beginning to give serious attention to this problem, both out of a sense of social responsibility and because it is essential to their labor relations. This is an area in which burgeoning technology of teaching machines and other techniques of self-instruction — for example, using computers — may play a specially significant role.

Although educational obsolescence is most serious for the man with little education, its impact is falling to an increasing degree on the engineering profession itself, and this is likely to be even more true in the future. Dr. T. E. Stelson of the Carnegie Institute of Technology in a paper, "Education for Oblivion," states the problem very eloquently:

> The decline in value or obsolescence of engineering personnel may likely become an increasingly serious problem in modern technology unless professional societies, employers, and schools recognize its importance and develop suitable remedies . . . unless a graduate of 10 years ago has systematically spent about 10% of his time extending his knowledge beyond the level of development achieved in his collegiate training, he will not have value in excess of that of a new graduate. . . . Faculty members who concentrate on teaching methods and educational concepts, no matter how commendable these aims are, are soon worthless because of technical obsolescence. In 10 years they cannot teach graduate courses — in 20 years they cannot teach advanced undergraduate courses.

We would disagree with Dr. Stelson only in that we believe he has understated the rate of obsolescence. This particular problem is one that deserves urgent attention by the Engineering Societies. Here again self-instruction techniques have a large role to play. Possibly the Engineering Societies should provide special self-instruction materials for members wishing to upgrade their education in their spare time. Something more may be needed than the bewildering and massive outpouring of technical literature, which gives information but not the framework and new perspective in which the obsolescent engineer can absorb it.

Perhaps the import of this section can be summarized by saying that to an increasing degree the business corporation and the professional society must regard education as an important part of their respective missions. Just as a corporation must continually improve its capital equipment, so must it develop its human resources. This has long been recognized as a necessity in the development of management. It has also been recognized in most large companies in connection with starting technical employees by the establishment of regular company training programs, and outside educational opportunities. It is less commonly recognized for more senior technical personnel and for workers below the professional level, though there is much increased use of training programs and courses of a specific character in the more progressive and fast growing industries.

MILITARY TECHNICAL TRAINING

As military systems become ever more complex and sophisticated, war itself becomes more demanding on human skills and education. If all the military weapons systems now under development were to reach operational use, the requirement for maintenance personnel, plus technical operating personnel, would far exceed the capabilities of the armed services to provide the necessary training and instruction by current methods. Despite increased automation our future defense posture is likely to become increasingly limited by the capabilities of human beings. This seems almost self-evident in the case of the sophisticated forms of warfare such as missile systems, aircraft, ships, or submarines. It is less obviously, but equally, true for the more limited forms of warfare, especially the type of non-overt warfare favored by our present antagonists. Counterguerrilla warfare, for example, requires the most sophisticated knowledge of geography, politics, government, economics, and languages as well as technology. In this form of warfare heavy reliance must be placed on small groups or even individuals acting independently. It is clear that the level of training demanded is of the highest order.

All of these requirements justify a hard look at military training technology and the use of all the educational aids possible. They involve education in both the intellectual and the psychological sense. Just as the technical training received by men in the armed services has resulted in great benefits to the civilian economy, so should lessons learned from experimentation with military educational methods be often adaptable in part to civilian education. This is far from the case now. Training in the military is poor and inefficient except in selected areas. Yet military education, especially at the level of the enlisted man, provides a laboratory in which new educational methods can be tried out. For its own future, the military should be much more vigorous in supporting experimentation in training methods and devices and even in conducting pilot training experiments.

It is an unfortunate fact that in the past the pressure of national security has led us to introduce innovations both in technology and in political organization which we ought to adopt anyway, but which achieve public acceptance only in the military sphere. Perhaps the same may become true of education. We may regret this but should not fail to take advantage of it to the fullest.

UNDERDEVELOPED COUNTRIES

It is becoming clear that the major component of the technical aid problem is that of education and training. The striking characteristic of all underdeveloped areas is staggering under-utilization of human skills and capabilities. The few highly educated individuals are not much help. They usually become disoriented and isolated from their culture, and find in their native land a barren field for their talents and training because they are removed from the support of an educated population. In the industrialized nations of the world, education has changed from a quantitative to a qualitative problem. We have educated many followers but still too few leaders whose creative talents make the others fully effective. But in the underdeveloped countries we are educating a few leaders, but no followers. The problem is to provide masses of people with the elements of education, mostly just plain literacy. The West cannot hope to solve this problem in a quantitative sense. It does not possess the necessary manpower, and the nature of our culture makes the recruitment of such manpower very difficult. Attempts to recruit large numbers of teachers are likely to lead only to the sending of people to underdeveloped areas who reveal the weakest and most undesirable aspects of Western culture. Thus the West must proceed by demonstration and pilot projects and by teaching native teachers, principally by working with them in their own environment. The development of inexpensive educational aids from programmed instruction to phonographs represents a tremendous challenge to the engineer and the educator. This is an area in which we cannot afford to wait for perfection. Something is better than nothing, and we must be willing to experiment with relatively untried techniques and devices, at the same time setting up pilot experiments in the real environment which will enable us to minimize errors, especially those tiny but fatal errors which nobody foresees.

An important aspect of education in the underdeveloped areas is the interplay between the processes of education and the cultural and social environment from which the students come. This factor is often overlooked in the introduction of

Western education into new countries having a radically different cultural heritage from our own. With the introduction of universal education in the United States in the latter half of the nineteenth century, we found an educational process which was extraordinarily successful in overcoming the cultural inheritance of immigrant families of widely diverse backgrounds and making the new generation a part of American life. This process was often brutal and disrupting to the individuals involved. It was probably successful largely because the immigrant children were mixed in with native Americans and lived in the American environment among their contemporaries. However, it is a very different matter when education is being attempted within a wholly alien culture. Much research and experience are required to understand this problem and to adapt educational patterns to an existing culture without either destroying the integrity of the individual on the one hand or fatally degrading the quality of the education on the other. A part of the problem is to raise the goals of the individual without inducing him to wish to deny his heritage completely or exploit the personal advantage that his Western education may give him in a primitive society.

However, Africa and Asia are not the only places having underdeveloped areas. We have a very analogous problem within our own country — in the urban slums, in the rural South, among the remnants of the American Indians. In this country the problem is often associated with the economic and cultural inferiority of the Negro imposed by his history. The educational problem involved has much in common with that of underdeveloped areas, especially those whose cultural integrity has been destroyed by colonial domination. It is the problem of arousing in the culturally deprived individual a sense of his own potential for achievement, of inducing him to set his sights sufficiently above the horizons set by family environment and social background but in terms of achievement rather than status. This requires special techniques, unusual qualities of human sympathy, and good teaching informed by deep understanding of the psychological and sociological problems involved. Perhaps one characteristic of this situation is that it is a two-way street, the teacher having as much to learn as the student. It requires a much deeper knowledge of human development in its entirety than most teachers have or are capable of acquiring. Yet it constitutes one of the most exciting and challenging tasks which teachers can face.

EDUCATION FOR A COMPLEX WORLD

The psychological and social factors that are so apparent in the education of underprivileged people are to a degree important in all education. In order to meet the responsibilities of freedom in the modern world, all men and women need an enlargement of outlook which will allow them to understand and accept the solutions to problems in the world as it actually is. One reason that the United States exercised much more statesmanlike leadership in the world after World War II than after World War I was that its people were better prepared through education to accept from their leadership an initiative informed by a world rather than a purely national outlook. The advent of the European Common Market and the greater acceptance in the United States of the need to reduce world trade barriers are other examples of how a higher educational level can (but does not automatically) lead to support of more enlightened government policies.

Our rapid technical growth has left us with increasingly serious social maladjustments, often left over from an earlier era. The so-called "revolution of rising expectations" and the hypersensitive nationalism of former colonies are two examples. Education must be the main instrument for overcoming these maladjustments, unless they are to be effected by the methods of totalitarianism. Our hope for attaining any kind of stability in the highly technical world ahead of us must rest on the ability of everybody to look ahead and to understand the world and themselves, and adapt to both as well as to manipulate both. In a democratic society in which social innovation can occur only by consensus and consent, the educational level of the people must be such that they can understand and accept the changes that are necessary. This probably requires a much more sophisticated understanding of our economic, political, and social systems than children now acquire in school. Here again is a challenge to education, and to scholars in social and political science. As Ramo remarks (see p. 370 of reference 4) the only alternative is to "accept something approaching a robot-controlled world that consists largely of ignorant and uneducated masses who are slaves to a few individuals who push all the buttons on the machines." From this Ramo infers an "obligation of those of us who are engaged in the engineering side of modern science somehow to apply ourselves to help the process of education."

EDUCATION AS AN APPLIED SCIENCE

The history of most technology follows a definite pattern, to which education is probably no exception. Initially it is based on an art or craft acquired laboriously in each generation through apprenticeship and experience. It is characteristic of such a technology that it is founded on accumulated experience of what works surrounded by little understanding or even fairy tales as to why it works. The art is characterized by a high degree of perfection, difficulty in communication, and very slow progress. In the meantime science develops, but at first is able to illuminate only the most elementary and obvious parts of the art; at this stage science is looked on with contempt and ridicule by the practitioner because it seems to belabor the obvious. But finally science does catch up with the practitioner, and at this point the technology becomes an applied science. Since it is now explicitly described and understood, it can be more readily taught. At the same time it can be built upon rapidly because scientific understanding narrows the range of choices possible to achieve improvement and often permits big steps rather than minor advances. Eventually it permits formalization and ultimate mechanization of the processes of design or planning to meet defined conditions or specifications. This is the handbook stage, and most likely in the future the computer stage. We have watched this evolution in engineering and in medicine, and we shall soon see it in education.

At the present time teaching remains largely an art, based on experience and personal skill. Nevertheless, a science has been growing over the last thirty years. Mostly it has only been able to rationalize what the skilled teacher does instinctively; the science is just catching up with the art. However, a fresh wind is blowing over the educational enterprise. In the words of A. A. Lumsdaine (see reference 4, p. 563), it is the "concept that the processes of teaching and learning can be made an explicit subject matter of scientific study, on the basis of which a technology of instruction can be developed." This is the definition of an applied science. The relation between this new science and the teaching art is well expressed by the following paragraph, also from Lumsdaine:

> As we learn more about learning, teaching can become more and more an explicit technology which can itself be definitively taught. The belief that teaching is primarily an art with which the gifted teacher has to be born and which defies precise description thus gives way to the conviction that teaching consists of techniques and procedures which can, in large part, be made communicable or teachable. This is not to say that the talent of the superior teacher can be replaced. On the contrary, it seems clear that outstanding performance in teaching, as in any profession, is achieved only by those who, in addition to a firm grounding in a communicable technology bring to their practice a high degree of creativity and in-

spiration. This certainly must remain true in teaching as well as in medicine, law, architecture, engineering, physics, or musicianship. At the same time the highest achievements in any profession seem likely to be realized only when they build upon a well developed underlying technology.

What is described here, however, is not a technology in the conventional engineering sense, but a technology of human behavior. Indeed there is a real potential danger in overemphasis on the purely physical aspects of this technology — teaching machines and audio-visual aids — a degrading of the importance of educational content and the psychological processes involved in learning. It seems likely that the need for hardware will be met by existing commercial enterprise, provided this need is adequately expressed as a result of research and experimentation in our educational institutions. It is impossible to separate the engineering device from the instructional goals and the explicit subject matter content of the instruction for which it is designed. If engineers are to contribute to this area it will have to be as part of a team encompassing the coordinated efforts of teachers, students, psychologists, subject matter specialists, educational administrators, and engineers. Engineers will have to become intimately involved with schools of education and with school systems in helping to establish the requirements and specifications for the development of educational devices, and in assisting in the evaluation of such devices.

The principal fruit of the advances in the science of learning is the teaching machine. First suggested by S. L. Pressey in the 1920s, its modern development and popularization has been largely due to the work of Skinner (1954, see reference 4) and his students and associates. It is an application to the learning process of behavioristic psychology, and is based on the hypothesis that learned behavior is essentially similar in all organisms including man. The machine aspect is only incidental, a convenient and economical way of presenting the "program" which is the essence of the technique. The program is a sequence of decisions required of the student, each a slight advance over the previous one. The decisions are not of the "multiple choice" type but require definite action, such as supplying a missing word or number. The most important psychological principle involved is that of "reinforcement," the process by which correct behavior is rewarded, a form of feedback which tells the organism it has made the correct decision. In human beings, unlike lower organisms, the mere knowledge that a decision was correct provides adequate reward or reinforcement. By means of the learning program the student is led by very small steps from what he knows into new territory. It is obvious from this description that the heart of the teaching machine is the preparation of the program itself, which demands a high order of understanding and analysis of the subject matter to be taught and the goals of instruction. Thus the development of programs has an important by-product in improving the performance of the programmer as a classroom teacher.

The basic technological problem in programmed learning is the presentation of the program in a sufficiently practical and economical way so that it can be used and reused by many students. The preparation of the program requires high skill both in programming technique and in the subject to be programmed, but even after a good program has been prepared its reproduction for classroom use is very costly compared with ordinary textbooks. According to Lumsdaine it has been estimated that reproduction of a program for a two-semester high school physics course would cost as much as $50 per set, not to mention the cost of the display machine. Here there is considerable scope for the engineer in devising new techniques. Widespread use of teaching machines will bring down the cost of both of the machine and the program materials. For many business, industrial, and military training purposes rather elaborate programs designed for relatively few trainees may be economically justified. This should prove a fruitful ground for experimentation, as we have indicated earlier, both with programming methods and with devices for presentation. It may also serve as a means for covering development costs, and bringing down unit costs to the point where they can be economically introduced into regular school systems.

At the present time one of the greatest needs is for an efficient medium of information exchange, and for making trial programs in developmental stages readily available to other programming groups for experimental use. There is a close analogy here with the sharing of software which played such an important part in the rapid development of the use of high-speed computers in the 1950s. There is a need to formalize this exchange function through several central clearinghouse agencies. With respect to programs designed for the "upgrading" of engineers, which may ultimately become available, the engineering societies can play an important role. With respect to school programs the Office of Education can play an important role. Consideration might be given to the creation of an extension service aimed at bringing the fruits of program development and other forms of educational experimentation to individual school systems.

There can also be a strong interaction between physical technology and programming. Present emphasis is on fixed sequence programs, although some experimentation has been done with the scrambled page program in which alternate sequences are presented depending on the way a decision in a previous step in the sequence was made. It may be that much more extensive use could be made of variable sequence or branching programs, which would especially lend themselves to use with digital computers. S. Ramo (see reference 4, p. 379) has predicted the emergence of a new profession, that of the teaching engineer, who practices "that kind of engineering which is concerned with educational processes and with the design of the machines as well as the design of the material" (that is, program). Despite Ramo's prediction, it seems to us doubtful whether engineers can or should play the dominant role in the development of educational systems. Perhaps this is mainly a question of semantics. In the sense that teaching is becoming an applied science, combining a knowledge of behavioral science with a certain degree of familiarity with electronics, electromechanical systems, computers, optics, and other physical techniques, it might be regarded as a new branch of engineering. Nevertheless, we believe this profession should find its academic home in schools of education rather than in engineering schools. There should, however, be much more collaboration between engineering schools and schools of education in the study and development of educational systems and techniques.

There is an even more profound relationship between engineering and education than is implied by the above discussion. This exists at the theoretical level and is based on the analogy between the learning process and the behavior of adaptive control systems.[5] Behavior exhibiting some of the features of learning has already been simulated on digital computers. There is a useful analogy also between the whole process of education and an automatically controlled industrial process. The output or product is learned behavior — if you will, a highly complex and sophisticated system of conditioned responses. The raw material is information, and this must be processed through the student in such a way as to produce the desired learned behavior. It is also necessary that the process be controlled — that is, that the results be frequently measured and the learning process adjusted to minimize the discrepancy between the desired and achieved results. Just as many of the theoretical concepts of automatic control and information theory are being used to suggest analogies for the development of a theoretical

[5] See, for example, W. Ross Ashby, "Cybernetics Today and Its Future Contribution to the Engineering Sciences" (New York: Foundation for Instrumentation Education and Research, 1961).

science of management or decision-making,[6] so should the same theoretical framework be suggestive of fruitful approaches to the theory of learning and ultimately to the design and organization of education and training systems. Indeed, the potentials inherent in this general field of knowledge for the control of human behavior and human development are frighteningly great, for good or evil. We need to understand what they are, if only in self-defense.

Most work that has been done with teaching machines has been concerned with rather elementary instruction. However, the few relevant experiments which have been performed indicate that the principles may be equally applicable to many forms of advanced instruction, even at the graduate level and beyond. Some students believe that programmed instruction may be a valuable technique for teaching creative thinking, and for teaching such poorly defined subjects as the appreciation of music and literature. Intelligent experimentation in this area, without quackery, should be given encouragement.

The development of educational methods based on behavioral psychology is only one aspect of education as an applied science. The programmed learning technique is based exclusively on small incremental steps, the cultural analog of Darwinian evolution by natural selection. There is another aspect of the educational process which might be described as the sudden flash of insight, the leap forward in understanding, when everything suddenly falls into place after a long period of plodding hard work. This is a process that is known to take place under the influence of great teachers, but is much less well understood from the psychological viewpoint. There may be ways of exploring and fostering this phenomenon systematically.

A related area to the above, in which the engineer may have

[6] See, for example, J. W. Forrester, "The Impact of Feedback Control Concepts on the Management Sciences" (New York: Foundation for Instrumentation Education and Research, 1960).

an important contribution to make, is that of the understanding and description of the creative process, especially as it applies to technological innovation. This is a distinct psychological process which must be recognized as quite separate from the substantive subject matter of engineering. Understanding of the psychology of invention and discovery, like understanding of the learning process, is the first step toward an applied science of invention, which would permit it to be explicitly described and taught and hence applied much more extensively than at present.[7]

RECOMMENDATIONS

1. The most significant role for the federal government in support of education is the encouragement and support of educational research, both basic and applied, and as to both content and method. An important part of this activity should be pilot educational experiments on a sufficient scale to provide reliable evaluation of new techniques and to provide sufficient demonstration of success to lead to rapid adoption of the new methods.
2. Hand in hand with educational experimentation there should be an organization somewhat analogous to the Agricultural Extension Service aimed at disseminating the results of applied research to individual school systems, and for acceleration of the application of new knowledge to the educational process on the widest possible scale. Such a service is especially necessary in view of the fragmented and decentralized administration of our educational system.
3. Closer collaboration between engineering and educational

[7] W. J. J. Gordon, *Synectics, a New Method of Directing Creative Potential to the Solution of Technical and Theoretical Problems* (New York: Harper, 1961).

theorists may be fruitful in developing new theoretical insights into the educational process. This could be carried out effectively in universities having strong engineering schools and strong schools of education. At the same time collaboration between educators and engineers in the evaluation of possible modes of employment of educational devices would be highly desirable.

4. There is an urgent need for several centralized clearinghouses for exchange of information on new educational techniques and for exchange of teaching programs for evaluation.
5. The engineering profession should make greater efforts to describe and understand the creative process as it occurs in technological innovation and to apply the systematic knowledge thus gained in the education of engineers and even earlier in the educational process.
6. The Engineering Societies should take the lead in developing teaching materials for self-instruction to help older engineers in preparing for the rapid obsolescence of many engineering and design procedures which will take place in the next ten years as a result of the use of automatic data processing.
7. As in medicine, engineering can contribute to the greater efficiency of the nonteaching aspects of the educational system, including building layout, student record-keeping, and the planning of auxiliary services; engineers should play a larger role in research and instruction in educational administration.

TEN □ National Science Policy and Technology Transfer

The following chapter is a paper which was originally prepared as a dinner address at a conference on Technology Transfer and Innovation held in Washington under the auspices of the National Planning Association and the National Science Foundation during May 1966. The proceedings of the conference have appeared in NSF report 67–5 issued by the Office of Economic and Manpower Studies of NSF in 1967.

The paper owes a good deal to discussions with colleagues. It was written at a time when I was active as a member of a panel of the Department of Commerce, under the chairmanship of Dr. Richard Leghorn, concerned with innovation and economic growth, and several of the ideas discussed had currency in that panel. In addition, I profited greatly from many discussions with Mr. Daniel Shimshoni, whose thesis in the School of Public Administration at Harvard dealt with the history and characteristics of innovation in small science-based firms.

DEFINITIONS AND CONCEPTS

By national science policy in this paper I refer to the mechanisms, institutions, and operating principles through which federal resources are channeled into scientific and technological activities. In the division that has become customary in this area, it is principally "policy for science" rather than "science in policy."

The concept of "technology transfer" is more difficult to pin down. Our image of technology comes down to us from the nineteenth century, and we tend to view it primarily in terms of machines and physical tools — that is, hardware. However, today technology consists increasingly of "software" — that is, of the organization and systematization of ways of doing things and not merely the specifications for things themselves. Unless we view technology in this wider sense, our policies are likely to be based on an obsolete view of the transfer process. In this wider sense the principles of scientific management, when embodied in an organization or in a management system such as PERT, are just as much technology as a ballistic missile. Similarly, to the extent that social science knowledge is embodied in techniques such as market surveys, public opinion polls, educational tests, programming for teaching machines, or urban-planning procedures, it is technology. In short, technology consists of codified and reproducible ways of doing things derived from rational principles. It is necessary to take this broader view of technology because the process of technology transfer is similar whether we are talking about hardware or software, and because nonmaterial technology is the fastest growing, and is likely to increase in relative significance in the future. The fact that the program development costs for a modern computing system — that is, the programming to make it applicable at all as a system, not the programming for specific problems — often exceed the hardware development costs is only one concrete illustration of a universal trend.

Technology transfer is the process by which science and technology are diffused throughout human activity. Wherever systematic rational knowledge developed by one group or institution is embodied in a way of doing things by other institutions or groups, we have technology transfer. This can be either transfer from more basic scientific knowledge into technology, or adaptation of an existing technology to a new use. Technology transfer differs from ordinary scientific information transfer in the fact that to be really transferred it must be embodied in an actual operation of some kind.

KINDS OF TECHNOLOGY TRANSFER

I have already hinted at two different kinds of technology transfer, which might be called vertical and horizontal. Vertical transfer refers to the transfer of technology along the line from the more general to the more specific. In particular it includes the process by which new scientific knowledge is incorporated into technology, and by which a "state of the art" becomes embodied in a system, and by which the confluence of several different, and apparently unrelated technologies, leads to a new technology.

Horizontal transfer occurs through the adaptation of a technology from one application to another, possibly wholly unrelated to the first, for example, adaptation of a military aircraft to civilian air transport, adaptation of a military computer for a business or general-purpose scientific use, adaptation of a laboratory analytical instrument for on-line control of a chemical process, transfer of a laboratory technology such as high vacuum to an industrial technique. The last two examples may, perhaps, be thought of as having both a vertical and a horizontal component in the sense that the laboratory application may in a sense be considered as more "general" than the process application.

Other particular forms of technology transfer that are important include those involved in technical assistance to underdeveloped countries, the transformation of laboratory procedures into medical care, and the diffusion of military and space technologies into the general commercial market. These are all examples of what I have called horizontal transfer.

The organization of both government and industry is primarily designed for the vertical transfer of technology — that is, the progression of technology from science to final product. The present organization of research and development in both government and industry leans toward what might be called "vertical integration" with research at all levels of generality being supported within a single mission-oriented agency or within a single mission-oriented institution. The fact that about 85 percent of research in American universities supported by federal funds is through the mission-oriented agencies is one of many manifestations of this vertical integration. Vertical integration might be contrasted with a hypothetical "horizontal integration" in which research would be organized by disciplines or techniques, and transfer of technologies between different applications might be facilitated. One might also call these "instrumental" versus "functional" views of research.

Of course, the above categorization of research organization is grossly oversimplified, and no such dichotomy between instrumental and functional organizations could be created even if it were considered desirable. For example, the professional organizations of scientists and technologists tend to be grouped on a functional rather than on an instrumental basis, and thus to cut across the "missions" of industry and government. Similarly, the universities are organized on a disciplinary or functional basis rather than in terms of problems or missions. Thus both professional societies and universities tend to serve as a common meeting ground for a variety of missions and thereby provide a potential mechanism for horizontal technology transfer, especially at the more generalized levels. However, the level is often too general to be a primary source of specific innovations.

Even the universities cannot be thought of as entirely horizontal in organization. For example, university medical schools are vertically organized with basic science departments at one end and teaching hospitals at the other. They thus aim at a symbiosis of all the technologies and sciences necessary to medical care. Similarly, the traditional organization for agriculture in the land-grant universities has been vertical rather than horizontal, with a strong but persistent drive toward intellectual self-sufficiency within the context of a particular mission, ranging from basic biology to agricultural extension and marketing.

Within government we have also seen a trend toward functional, or horizontal, organization. In some degree, for example, the Atomic Energy Commission (AEC) and the National Aeronautics and Space Administration (NASA) are mixed organizations in that they are responsible for technologies with several different missions, and they have a definite legislative charter to discover and develop new missions in the context of a broad area of technology. Proposed extension of the work of the National Science Foundation (NSF) to include applied research may be a step toward a horizontal integration extending across all of science. Nevertheless, NASA and AEC remain vertically organized in the sense that they are designed to optimize the transfer of technology between different levels of generality from basic research to prototypes. It is no accident that both of these agencies seem much more preoccupied with the horizontal transfer of technology than do the older government science agencies, such as Agriculture. NASA, for example, is much concerned with spin-off from space technology and attempts to induce its contractors to seek out civil applications for their space-related developments. The AEC from its beginnings, and especially since the amendment of the Atomic Energy Act in 1954, has had an aggressive program of finding new uses for atomic energy and encouraging their adoption in industry, agriculture, and medicine. Both agencies have a well organized and supported technical information program aimed at maximum diffusion of their respective technologies.

Even industry has tended to compromise with the vertical-type structure which it originally developed and exploited so successfully. In fact, the growth of a research and development industry in its own right is in some measure a symptom of this change. Indeed, we now find a few industries which are beginning to define their mission in terms of horizontal technological transfer. Such industries declare as a matter of policy that their primary goal is to grow through the creation of new businesses derived from a few central technological themes rather than through the application of technology to predetermined goals or needs of society.

Thus, as we examine our R and D intitutions more closely

we find an ambivalence toward technology transfer rather than a singleminded devotion to either the vertical or the horizontal concept. I like to believe that this ambivalence is healthy, that its tensions in themselves produce a drive toward innovation. The process of technological diffusion is inherently two-dimensional, or perhaps more accurately, multidimensional. The great diversity of our technical institutions tends to ensure that diffusion occurs preferentially along certain dimensions in certain types of institutions. By optimizing each institution for a particular dimension we probably achieve a faster rate of diffusion than from trying to achieve an optimal compromise for all dimensions within every institution.

ROLE AND ATTITUDE OF GOVERNMENT

Since the federal government supports over two thirds of the scientific and technological activity in the nation, the way in which it does this may have a large influence on the process of technology transfer, whether it be in the vertical or the horizontal sense. I use the word "may" in preference to a more positive term, because I think the very diversity and variety of federal activities in science and technology may reduce its impact in any one direction. One might liken the impact of federal policy to the pressure of a gas — the composite impact of a variety of activities, rather than the driving force of a jet pushing in one direction. It has become fashionable to deride the randomness, or apparent randomness, of federal science policy, and to assume that a more directed impact would achieve our goals with less expenditure of effort and resources. In theory this may be true, but the theory presupposes a consensus regarding goals for which our pluralistic society is not notable. In the absence of such a consensus the best way to optimize progress may be to ensure the release of the diffuse energies of our society in as many directions as our resources permit, to maximize the technical choices ultimately open to us. This is, perhaps, an appeal to the old "invisible hand" argument of the classical economists in a somewhat different guise, but as in economics it still does have a certain wisdom. Certainly, there are circumstances in which little management may be superior to bad management, and in the present state of our knowledge of science policy, this is probably a realistic description of the alternatives with which we are faced.

Our institutional arrangements for technology transfer seem to be best designed in those areas where there is a clear consensus on federal responsibility. Within the last few years widespread concern has been expressed that our cold-war preoccupation with military and space technology has deprived the civilian economy of its sources of technological innovation and starved the public sector in areas related to the improvement and preservation of the human environment. It is somewhat ironic that at the same time that the United States is going through this soul-searching regarding the social effects of its excessive concentration on the most advanced and esoteric technologies, the Europeans are looking with awe and apprehension at what they see as the tremendous competitive economic advantage conferred on the United States by its massive investments in space, defense, nuclear energy, and health technologies. To some extent these two concerns are not as paradoxical or contradictory as they seem. Both represent partial truths. Foreign trade, at least among the industrialized nations, is disproportionately concentrated in science-intensive products and services. Thus, in principle, it would be quite possible to improve our foreign trade position while we lag in the application of science and technology to our total domestic well-being. This may be to some degree our present situation.

Our failure to apply science in the alleviation of pollution, the solution of urban transport problems, the stimulation of economic growth, or the rationalization of medical services is not a failure of science policy *per se,* but rather a failure to develop the institutions that could apply the necessary technologies when available, and this failure in turn stems from our inability to achieve a national consensus, at least until very recently, on the degree of federal responsibility in achieving these public goals.

Implicit in much of the discussion of the federal government in relation to civilian technology is an assumption that direct subsidy of research and development is the only mechanism by which government stimulates technology transfer. This assumption arises from experience with defense and space technology, where in fact direct government subsidy has played a major role. One reason such subsidy has been successful in these fields, however, is that they do not require a widespread consensus, other than in very general terms. It is true that Congress must vote the funds for a space program, and needs evidence of public support to do so. But, by and large, the space program affects people only in general and does not have widespread impact in detail. Only a few people have to be convinced of the detailed technological procedures necessary to get a man to the moon. Space and defense are spectator sports. The public and the Congress feel, rightly I think, that they can afford to leave the detailed decisions to the technologists, and that these decisions will have only a generalized impact on their daily lives. On the other hand, when it comes to matters like urban transportation or pollution control, particular interests are heavily involved and differentially affected. Public support for the general ends is insufficient; there has to be widespread consensus as well on the means because they usually involve the sacrifice of particular interests to the general good. Even the technical choices in R and D affect private and local interests sufficiently so that pressures are generated to avoid developing the knowledge on which effective public action might be based. The selection of alternative technical routes involves much besides technical considerations. It involves anticipation of the response of the people, and often testing of these responses at intermediate stages with feedback into the technology itself. In the last analysis, will people choose to take advantage of urban rail rapid transit if it is available? Are our tools of technical-economic analysis sufficiently refined to really justify the large gambles that are often involved in the decision to build a new rail transportation system, or a new system of waste disposal? In such cases the better governmental role may be to provide the framework of information, incentives, and underlying general technology, but to leave the initiative for development and application to private and local institutions. If government is to adopt this philosophy, how is it to demonstrate or measure success? How is it to decide in practice where the line between public and private initiative is to be drawn? It is this type of issue which often frustrates the consensus necessary for progress. In the area of "public goods" there is a growing feeling that the more public policy can achieve its goals by indirection rather than by direct subsidy of R and D, the more likely is an optimal solution to be achieved. Thus to the degree to which the public itself can be brought into the decision between technical alternatives through quasi-market mechanisms, the ultimate accommodation between public and private interests is likely to be more acceptable and workable.

There is another area in which, until recently, public policy has been quite inactive, and this has to do with the influence of federal science policy on the innovative climate in the private sector. Only recently has the Department of Commerce taken some leadership in this field. However, this leadership has yet to receive real recognition at the national or political level. In the first place, there does not appear as yet to be a national consensus that a high rate of innovation in the private economy is a desirable national goal in itself, or, if so, a goal which should be of concern to public policy. Here again we have a case where the adverse effects on particular interests may have more political horsepower than the general benefits. The costs of social change resulting from technological progress are too often borne by those elements of society least able to sustain them, and we have not developed the political, social, or economic mechanisms for distributing the costs of technical progress sufficiently equitably so that its net desirability can become more obvious. The report of the Automation Commission seems to have made a beginning at facing this problem honestly and frankly, and it is a heartening measure of the long way we have come in our social and economic attitudes in the last fifteen years that a group representing such diverse interests in our society could arrive at such a large measure of consensus as did that commission.

Coming to national science policy, however, it is clear that there is still an important element of policy missing. There has been no succinct and clear-cut statement of national policy to the effect that the federal R and D investment, and the federal policy toward technology in general, should be designed so as to achieve the maximum possible beneficial impact to the economy consistent with other goals. There is precedent for such a statement in the President's memorandum to the agencies concerning the impact of federal R and D on higher education. This is, perhaps, the clearest statement in public policy that the impact of federally supported R and D extends beyond the immediate goals for which it is undertaken. In this sense it is a recognition that the fully vertical integration of federal R and D constitutes an insufficient basis for policy. It seems to me that the time may be here when a similar national policy statement may be desirable concerning the impact not only of federal R and D on the private economy, but also of other federal policies on the general climate for innovation. Such a statement might begin with an affirmation that a rapid rate of technological innovation is an indispensable ingredient of economic growth, and that henceforth federal agencies engaged in the support or conduct of R and D should attempt to shape their policies with due attention to their economic impact, including particularly the horizontal transfer of technology from the immediate purposes of the agency to other purposes and to the civilian economy. Such a declaration should serve to clarify agency thinking and, over a long period, gradually reshape the criteria on which R and D decisions, especially those in the applied area, are made. A number of agencies are, of course, devoting considerable attention to this problem. NASA, in particular, has always stressed the importance of "fallout" from its programs. In its contracts with universities it has attempted to require a responsibility for coupling to the regional economy, and it has sponsored detailed studies at M.I.T. and elsewhere of the origins of science-based industry as spin-off from federally supported R and D institutions. The Defense Department has also sponsored a number of studies of science-based industry, and has begun to give more serious attention to the regional impact of its R and D programs, stimulated in part by its concern for the possible effects of cutbacks in defense expenditures. These and other activities suggest that the time may be ripe for a more coherent statement of policy. Although the government is already fairly well organized for vertical transfer, the new statement of policy would emphasize the stimulation of horizontal transfer as well — that is, the transfer or diffusion of similar means to a variety of ends.

Some of the ways in which federal R and D policy influences technology transfer may be listed as follows:

1. Through the support of basic and academic research by the mission-oriented agencies.
2. Through the mix of institutions and organizations to which support is given, including extramural and intramural institutions and, among extramural performers, the balance between industry, federal contract centers, nonprofit institutes, and universities.

3. Through the selection of mechanisms for support, especially project versus institutional funding.
4. Through the types of controls exerted over supported institutions, including such matters as conflict of interest, outside consulting, segregation of government and non-government work, reporting requirements, and so on.
5. Through its policies for the diffusion and indexing of scientific and technical information.
6. Through patent and copyright policy as applied to sponsored research and development.
7. Through the type of funding instruments used, and the way in which problems are defined.
8. Through its use of external advisers from industry and universities.
9. Through the geographical distribution of R and D funds and the criteria used in the selection of contractors and grantees.

Let me give a few illustrations.

The classic argument for the support of basic research in universities by mission-oriented agencies, rather than by a single source such as NSF, is that the administrative environment provided by the mission helps to ensure more rapid identification of new opportunities for mission technology emerging from science, and at the same time gives the agency an opportunity to bias the research in particular directions having greatest potential payoff for its missions. In addition, by sponsoring research in institutions that are at arm's length from the short-term goals of the agency, there is a greater probability that the work will have other payoffs than the ones with which the agency is immediately concerned. Thus the argument touches on both the vertical and the horizontal transfer of technology in the sense I have used previously. The administrative connection with the sponsor facilitates vertical transfer, while the remoteness favors horizontal transfer. Intramural research support favors vertical transfer, while extramural support favors horizontal transfer but also gives access to a greater depth of science and technology for vertical transfer. Although all these arguments have considerable plausibility, they are for the most part undocumented, and such documentation as exists is anecdotal rather than quantitative. There is serious need for more critical scholarship in this area. The Defense Department has made one such attempt in its "Project Hindsight." The so-called Westheimer Report of the National Academy of Sciences, concerned with opportunities and needs of basic chemistry, broke new ground in this field of study through using literature references to trace the basic research origins of new developments in chemical technology. There has also been a recent upsurge of scholarly interest in tracing the origins and history of science-based industry.

A number of federal agencies in the space, defense, and nuclear energy fields have allowed an item for "independent research" in the indirect cost pool for large procurement contracts. The primary purpose of this policy was to encourage industrial research related to federal business through recognizing it as a legitimate cost of doing business. However, it may have an important indirect effect in encouraging technology transfer to the civilian economy. If properly administered, such research can be carried out in intimate contact with the company's normal activities, and so has a higher probability of horizontal transfer than research aimed directly at defense objectives.

Through sponsoring research and development in a wide mix of institutions the government ensures a variety of institutional environments for research, thereby increasing the probability of technology transfer. Thus basic and applied research carried on in an environment that is closely tied to a heavily projectized work will have a high probability of rapid application, but will tend to be steered within the context of a narrow mission. On the other hand, research carried out in an academic environment will have a lower probability of being identified for particular applications, but because of the rapid mobility of people through the academic system, the transfer of technology to a variety of purposes occurs through the movement and diffusion of people trained in new technologies, especially those connected with basic research. The academic environment tends to favor "serendipity" or accidental association in the process of technology transfer, but is less responsive to problem solving.

Many of the types of administrative controls exercised by government agencies in contracting with industry may tend to favor isolation of government from commercial work. This isolation is often encouraged by industry itself because the challenging, science-limited technology of defense, space, and health tends to be incompatible with the cost-limited technology of the commercial market. Nevertheless, this isolation may have serious long-range consequences for innovative activity in the commercial sector, and federal policy at the least should discourage rather than actively encourage it.

Since new technology is usually incorporated in new products, services, or production processes any government action that tends to increase the demand for such products and services has the effect of stimulating innovation. Thus heavy federal support of R and D in all sectors has not only produced an outpouring of direct research results and of planned innovations but has also created a commercial demand for highly sophisticated instrumentation, including such things as analytical instruments, electronic measuring instruments, and scientific computers, and also of technical services or "software." This demand has often been the first step in a process of technological diffusion in which products developed to meet a highly specialized demand from the laboratory or the military have gradually found their way into the general economy. The commercial-type demand generated by research use has financed the most expensive part of the "learning curve" and permitted the perfection of products in respect to cost and reliability to the point where they could penetrate wider and wider markets. Although the scientific instrument market is small in terms of fraction of GNP, its true economic significance may be very large because there are many important products and production processes which would be impossible without this specialized instrumentation. For example, it has been stated that the widespread introduction of the basic oxygen process in steelmaking would not have been practical without the availability of modern process control instrumentation, which is nevertheless only a small part of the total cost of production. Similarly, the design of modern commercial jet aircraft would be impractical without high-speed digital computers. The input-output studies of Leontief and his associates comparing 1947 and 1958 matrices already reveal the sharp increase in "general input" depending for the most part on science and technology across almost the whole of our production economy. Indeed, it seems possible that the impact of federal R and D may have more economic significance in terms of the types of demand it creates for sophisticated new instrumentation technologies and technical services than it does for the direct results of R and D, an intriguing possibility that would merit further study.

The federal government has a strong influence on innovative activity through the ways in which it defines its R and D problems and the way it determines the specifications for the products and services which it procures from the civilian economy. A very narrow definition of its problems prior to contact with the outside world tends to have an inhibiting effect on innovation and on the horizontal transfer of technology between different purposes. To the extent to which Congress tries to legislate innovation — what I like to call invention by definition — it probably defeats its own goals. Federal procurement of R and D may legitimately define the goals to be achieved, but these should be specified in functional terms, not in terms of the methods to be used in achieving them. Again a clear statement of national policy would clear the air in this domain.

MECHANISMS FOR TECHNOLOGY TRANSFER

The mechanisms of technology transfer in the broad sense in which I have been using it in this paper may be listed as follows:

1. The movement of people between different fields of science and technology and from science into technology.
2. Entrepreneurial activity in the broad sense, that is, the spin-off of new missions or enterprises from existing organizations.
3. The scientific and technological literature; and the activities of professional societies.
4. Interaction between the supplier and the customer, or more generally the developer and the user.
5. Programs of training and education.
6. Consulting and advisory activities.
7. Patents and trade in know-how.
8. Marketing and applications engineering.
9. Accidental personal contacts.
10. Technical meetings.

I should like to comment briefly on each of these mechanisms, with special attention to the role of science policy.

The movement of people is one of the most effective forms of technology transfer, and case studies of specific new technologies have begun to document this. The types of mobility follow the two modes, vertical and horizontal transfer, which we have already mentioned. Vertical transfer thus occurs significantly within the firm through the movement of basic research people into applied research and operations, and less frequently through movement of technologists into basic research. Industrial laboratories often recruit many of their technical people initially from university research in the hope of interesting a certain proportion of them in moving, along with the newest technologies of basic research, into new applications. Such movement of people is facilitated when a single institution, such as a company, spans the full range of technical activity from basic research to operations. However, such movement of people is seldom successful unless it can be induced to occur spontaneously. If applied research simply becomes the limbo to which the unsuccessful basic researcher is consigned, then the quality of applied research will suffer, even though it is true that *some* valuable applied research can be performed by people who would not be successful at basic research. The good mission-oriented laboratory is one that provides an environment in which a certain proportion of basic research people will find themselves challenged by applied problems or will come up with ideas for application which they will wish to follow at least partway through to final use. Too often the government laboratories tend to place people in niches where they stay for the duration of their careers. Because of difficulties in recruitment, especially at the higher levels, government laboratories have got into the habit of thinking of turnover as a sign of ill health, and a high retention rate as a sign of vigor. On the contrary, federally supported research organizations should be encouraged to look upon mobility as a good thing and to regard the achievements of their alumni with as much pride as the achievements of their staff. There should ideally be more recruitment between government laboratories and greater interchange between scientists in the laboratory and headquarters staffs.

At the same time university researchers should come more to accept the idea that not all the Ph.D.'s trained in basic research ought to stay or be supported in basic research for the balance of their careers. This fact is much better accepted in some fields than others. For example, in chemistry only 30 percent of the Ph.D.'s produced are retained in the universities, and this contributes toward what I feel is a healthier attitude toward applied research in chemistry departments and at the same time an innovative and forward-looking industry.

The next point, entrepreneurship, is closely related to the movement of people. Many organizations are beginning to recognize that spin-off is a desirable phenomenon and to take pride in and encourage it. Spin-off includes not only the creation of new enterprises but also the creation of separate budgeting centers within organizations which market their services on a quasi-self-sustaining basis.

In this connection I might remark that the entrepreneurial character of the project grant system in the support of basic research has some merit. Ideas and performance are "sold" in a kind of intellectual marketplace, where price is not a major criterion, but where other performance criteria should be.

Entrepreneurship encourages technology transfer because people seek new outlets for technology, potential solutions seeking out new problems as well as problems seeking new solutions. Entrepreneurship thus fosters horizontal technology transfer.

Research institutions provide a home, a form of interim personal security, for would-be technical entrepreneurs while they are in their starting stages. There is a need for large federal research institutions, especially, to provide an environment that is hospitable to this type of enterprise. It is true that this makes the problems of protecting the public interest more tricky, but if such enterprise is recognized by public policy as desirable, as a new goal for science policy, I think the means can be found to ensure proper protection of the public interest.

Next, as regards publications and information activities, studies indicate that this is not as important in innovative processes as the direct interaction of people. The information explosion is a serious problem, though I believe it has been oversold. The problem is not so much access to information as it is identification of relevant information. Present information systems are not well matched to the hierarchical and associative structure of the human mind. Another problem is how the scientist or technologist comes into contact with the information he does not know he needs. Case histories of recent innovations in science-based technology underline the importance of random contacts. Innovation often begins with the recognition of the relation between pieces of information hitherto regarded as unconnected. Even slight improvements over a random process could pay large dividends in this situation.

Next I come to the supplier-user interaction. The sophisticated buyer often plays a large role in fostering innovations and technology transfer. The pattern of diffusion of technology often follows a course from a military or laboratory market toward a consumer market, with a producer goods market as an intermediate stage. At one extreme lies the scientist who designs and builds his own laboratory instruments; in this case the supplier and the user are the same man. At the other extreme the product from the user's standpoint is simply a black box that performs certain functions, but whose inner working he does not understand or care to understand. But at every stage the supplier and the user collaborate in some sense, and the collaboration at one stage prepares the product or service for acceptance at the next lower stage of customer sophistication. At the beginning of this process, the performance of the product is everything, but as we progress through the "product cycle," cost, reliability, and foolproof operation assume increasing relative importance as the product or service becomes standardized.

What is significant here for government policy? The government should actively encourage transfer of responsibilities to external or independent suppliers as early as possible. It should encourage innovation not only by direct subsidy of R and D but by creating an "induced demand" for sophisticated products and acting as a sophisticated buyer to set performance criteria and standards. When purchasing R and D services directly, it should specify the results desired in as broad and general terms as possible consistent with its defined objectives. Intramural research activity should be slanted as much toward enhancing the government's capability as a sophisticated buyer of innovations as toward creating innovations.

As regards the scientific and technological literature, studies of innovation suggest that the literature by itself is not a major source of technology transfer, though there are exceptions to this. The literature is probably more important in providing a source of knowledge for developing and improving an invention rather than for the invention itself. The difficulty is that the very volume of scientific literature makes the "noise level" of unwanted or unneeded information too high to be useful, until the innovator has clearly defined his information needs, and this usually happens only well after the initial invention. Nevertheless, from the standpoint of government policy in fostering technology transfer, wide dissemination of general technical information resulting from government-supported work should be an important objective of policy, particularly through encouraging publication in recognized journals that are widely diffused in the technical community.

In regard to training and education programs, both government and industry in recent years have begun to recognize the importance of continuing education in the maintenance and upgrading of its human capital. Nowhere is this more important than in science and technology. There is a gratifying increase in the use of sabbaticals and training leaves among professional and even semiprofessional people. There is also need for an opposite flow which helps expose people trained in basic research to technological needs. A fruitful way of achieving this may be through the provision of temporary postdoctoral positions for Ph.D.'s in industry and government laboratories. At least some of these positions should be designed specifically with the idea of promoting technology transfer. They should be aimed at demonstrating the challenge of applied work or of exposing younger university scientists to new technologies that are especially strong in government laboratories.

Studies seem to indicate that specific technology transfer to industry through graduate students is not very important. The value of the movement of students from universities to mission-oriented institutions seems to lie more in the general sophistication and open-minded approach to problems which they bring, rather than in highly specific skills. The most enlightened industrial laboratories often recognize this fact by assigning fresh Ph.D.'s to problems wholly outside the field of their thesis work. Particularly in its early days, solid-state electronics in industry benefited greatly from the work of young scientists whose original training was in nuclear physics or cosmic rays.

Consulting and advisory activities provide a way of involving basic scientists in applied areas without a full-time commitment on their part. This is an important way of maintaining a contact between the basic research community and the mission needs of government and industry. The Department of Defense has benefited particularly from this kind of involvement, and through the bringing together into "study groups" of people of diverse technical backgrounds and experience to deal with a coherent set of problems.

Patents seem to be declining in importance as a means of technology transfer. Studies of science-based industry indicate that, apart from a few spectacular exceptions like Xerox, patents are not a significant basis of new enterprises in sophisticated technology. More often patents are taken out to protect against exclusion by others, or to provide a trading base for exchange of know-how with other firms. Some observers have expressed considerable concern over the decline of patents relative to R and D expenditures. Others express little concern and ascribe the declining importance of patents to the fact that broad capability for rapid exploitation and development of new inventions as they appear is far more valuable than the inventions themselves. Nevertheless, the importance of patents varies greatly from industry to industry.

An interesting question regarding patents is why the source of financing for an invention should be the sole basis for equity interest in the patient. Should an inventor even be permitted to assign his entire interest in an invention to the agency or institution which financed his work, or should he retain a small residual interest in his own name which he cannot dispose of? The argument advanced against this is that it tends to undermine the team spirit so important for successful organized R and D, but one wonders whether offsetting advantages are not sacrificed. Would not industry be in a stronger position in arguing for more generous federal policy regarding the retention of patent rights arising from government-supported R and D if a residual equity were retained by the individual inventor? There are many problems in such a suggestion, not the least of which is that of assessing the economic value of a patent which is utilized wholly within the firm. Still it seems to merit greater attention than it has received.

Marketing and its companion activity, applications engineering, cannot be discounted as a factor in technology transfer. We ordinarily tend to think of marketing in the sense of the private business market, but an analogous process may be equally important for innovation and technology transfer wholly within the public sector. It is an aspect of the supplier-user interaction to which I referred earlier, but stresses more the role of the innovator in introducing and adapting his innovation for the customer. When government enters areas of applied research, it too often tends to neglect the marketing aspect of the innovation process. Only in agricultural research has marketing of technology and applications engineering been clearly recognized in the institutionalization of the total innovative effort. One would like to see the active "marketing" of new technology between federal agencies and missions recognized as an important principle of federal science policy. By this I mean marketing of technology itself, not merely of R and D services. Perhaps the major broad spectrum federal research centers should have a specific responsibility for marketing new technology to agencies throughout the federal structure. An organization like the Bureau of Standards might actually have applications engineering groups whose function would be actively to search out ways of applying technologies in which the Bureau is expert to the mission needs of other federal agencies. As these groups found successful applications, some of their members might actually transfer to the agency in which the technology found application, thus contributing toward a desirable mobility of technical people within the governmental structure. A process of somewhat this character actually took place when NASA was created out of the old NACA. At that time many of the people from several naval laboratories who had become skilled in the new space technologies transferred to the NASA organization.

The new role that may be assigned to NSF also lends itself to a marketing approach for more basic types of research. Should not NSF be more active in "selling" support for certain areas of basic science which appear ripe for technological exploitation to other federal agencies? An instance in which it has already performed this function quite successfully is that of weather modification research. Here a number of other federal agencies are now anxious for the chance to use on a larger scale the promising research results which have been developed under NSF support. NASA has done somewhat the same thing in the field of weather and communications satellites. Both of these programs in NASA only became successful when a marketing-type approach was adopted and the agency was prepared to work as a vendor attempting to adapt what it had to what the customer agency wanted rather than what NASA thought he ought to want. Similar opportunities may lie before NSF in other areas, for example, information retrieval, ecology, genetic manipulation, and so on.

Accidental personal contacts have played an important part in the stimulation of innovation. There is not much that can be said about this in relation to science policy, except to underline the fact that much innovation takes place in an unplanned way, and policy needs to be sufficiently flexible to recognize this fact.

Finally there is the matter of technical meetings. Everybody today complains about the proliferation of specialized meetings, and the proportion of time spent by the scientist and the technologist in just "communications." This seems inevitable as the volume of scientific information grows exponentially. Probably meetings are more important for the random technical contacts that they generate than for the technical papers presented. The technical papers provide a respectable intellectual backdrop against which the real drama of technological transfer is played out in the corridors and the bars. The government probably makes a very large investment in technological transfer through paying the expenses of scientists to go to meetings, directly and indirectly. However, the government can play an important more specific role in organizing and supporting smaller meetings which cut across disciplinary lines and bring together basic scientists and technologists in groups with a problem-oriented agenda. The agenda must be broad enough to appeal to the scientist but specific enough to elicit information of potential value for technological applications. The Air Force has been especially forward looking in stimulating and organizing meetings of this character.

SUMMARY AS RELATED TO FEDERAL SCIENCE POLICY

The principal conclusions of this paper, insofar as they relate to national science policy, may be summarized as follows:

1. A high rate of technological change in all segments of the economy, public and private, is desirable for economic

growth and public welfare and should be a consideration in the formulation of a national science policy.

2. Thus federal policy should include a purpose of fostering the rapid diffusion of the technology induced by government research and development, and consideration of the economic impact of government R and D.
3. Government research policy should avoid encouraging the development of completely self-contained capabilities within government-supported mission-oriented research institutions, but rather should encourage development of independent suppliers who market their technical services and products to the parent organization and ultimately to other users inside and outside the government.
4. Wherever feasible, government should attempt to stimulate technological innovation by means of "induced demand" rather than by direct support of R and D or purchase of specifically defined R and D results.
5. Government support of any institution or class of institutions should encourage rather than inhibit mobility of people from one type of institution to another. For example, not all Ph.D.'s trained in basic research should expect to be provided with careers in basic research through government support. Civil Service policies should deliberately foster mobility of technical personnel between federal laboratories and agencies.
6. In order to foster horizontal in addition to vertical transfer of technology large broad spectrum federal research institutions should regard it as part of their function to market new technologies to other federal agencies and missions and to institutions in the private sector.

ELEVEN □ Applied Research: Definitions, Concepts, Themes

During the latter part of 1965, the Committee on Science and Public Policy of the National Academy of Sciences agreed with the House Science and Astronautics Committee to conduct a study on applied science along the same lines as its earlier study of basic science which led to the report "Basic Research and National Goals." The organization of this study and the selection of participants proved a considerably more difficult problem than the earlier effort. Applied research is more complex and diverse in its goals, its standards, and its style than is basic research. It can range from pure empiricism to abstract theory, from the highly particular to the very general, from a highly individualistic enterprise to a highly organized and programmed team effort. At one end of the spectrum, applied research is barely distinguishable in its style and method from the purest type of basic research; at the other end, it is best characterized as enlightened tinkering. Furthermore, all the styles and methods in this spectrum may enter at one time or another into a complete process of innovation.

As chairman of the Committee on Science and Public Policy, I began the study with an extensive correspondence during the fall of 1965 and the spring of 1966 with numerous acquaintances in industry, government, and universities whom I knew to be engaged in or concerned with the application of science. After reviewing this correspondence, a meeting was held in the Academy in April 1966, to which several of the correspondents were invited, and at which the issues were discussed with the members of the committee. On the basis of the correspondence and meeting, the essayists for the congressional study were selected. At the same time, as my own contribution to the study, I attempted to summarize and interpret the information and opinions we had gleaned from the correspondence and meetings. The following essay is the result of this effort. Although I take full responsibility for the views and ideas expressed, the essay owes much of its inspiration to the correspondence and discussions described above, and even some of the phraseology is lifted bodily from letters or discussion summaries. The number of debts is too large to acknowledge individually here.

INTRODUCTION

This paper is partly a digest of ideas turned up by correspondents and developed at a preliminary meeting on April 24, 1966, and partly interpretation based on my own experience and observations. It is in ten sections as follows:

1. An interpretation of the distinctions between basic and applied research and a description of their interactions, emphasizing the continuity and indivisibility of the research and development process.
2. Some of the problems and difficulties in studying research as a process, noting especially the evolving nature of the relations between science and technology, which makes conclusions from historical studies of technology of limited relevance to current policy.
3. The relative roles of government, industry, and universities in applied research, including a fairly detailed discussion of the historical role of universities in applied research in the life sciences and of criteria for university involvement in applied problems.
4. The contemporary interaction between science and technology, stressing the growing role of technology in pure science and the feedback between pure science tools and industrial development, also the increasing trend toward conceptualization in technology.
5. The use of the attitudes and methods of fundamental research in dealing with applied problems.
6. Political and administrative decision-making about technology, the relative roles of the expert and the generalist, and the appropriate degree of societal direction of applied research.
7. Discussion of the university as the characteristic institution of basic science and the mission-oriented, multidisciplinary research institute — industry, government, or in-

dependent — as the characteristic institution of applied science. The concept of a "mission" is defined, and the attributes of successful mission-oriented laboratories are itemized and discussed.

8. The question of the status of applied research and applied scientists in the United States with a tentative sociological analysis of the "snobbery" that exists between pure and applied science. It is suggested that vigorous national pursuit of equality of educational opportunity may be one of the surest methods of producing more and better applied scientists, since these often come from "upward mobile" segments of the population.
9. The problem of assessing quality in applied research, suggesting that better documentation of applied science would have a positive influence on quality.
10. The seminal role of the technical entrepreneur in spearheading technological innovation.

1. DEFINITIONS AND CONCEPTS

In institutions whose missions include the application of research results to products or operations, the categorization of research into basic and applied is not too meaningful, and has little operational value. Industrial and government researchers feel particularly strongly on this point because, from the standpoint of research management, the basic-applied dichotomy tends to focus attention on the wrong issues. In fact, all research in a mission-oriented organization contributes or should contribute, however remotely in time, to the general objectives of the organization. On the other hand, there is clearly a spectrum of activities ranging from pure research on the one hand to technological development on the other, and to some extent one can locate research activities within this spectrum according to their "appliedness." This relates to two factors, the time scale on which the research is likely to find an application, and the specificity with which the domain of application can be foreseen or the work committed at the time the research is undertaken. The shorter the time horizon and the more evident the area of potential application, the more "applied" the research. Furthermore, there can be a perfectly viable difference in viewpoint between the research worker and his sponsor. Research that may be viewed as quite fundamental by the performing scientist may be seen as definitely applied and may fit into a coherent pattern of related work from the standpoint of the sponsoring organization or agency. The scientist may see his own work in a matrix of interconnections entirely different from that in which the sponsor sees it. Furthermore, in a well-coordinated group of scientists, success in a particular line of applied research may greatly expand the possibilities for basic research. For example, when a new area of development opens up, an important benefit of intensified basic research related to this area is the indirect one of maintenance of technical standards, and the introduction or perfection of new intellectual and experimental tools that might not otherwise be justified. It is not necessary to control the direction of the efforts of the individual research man in order to realize these benefits.

One may discern in the literature two views of the processes of invention and innovation. One is the completely rational view, characterized by such terms as the "management of research" and "the organization of invention" or "cost-benefit analysis." It views innovation as a process that can be completely planned and that is designed to "convert the essential resources of human talents, physical facilities, money, and knowledge into marketable goods and services." [1] The other view, while conceding the need for planning and foresight, stresses the nonrational and fortuitous elements in the innovation process, the

[1] Quoted from John F. Mee (October 1964) in D. F. Schon, *Technology and Change* (Delacorte Press, New York, 1967), p. 3.

fact that "once a process of technical development has begun, it does not usually move in a straight line, according to plan, but makes unexpected twists and turns." It lays emphasis on the "unexpected boundaries of need and technology," on the fact that "invention" often "consists in carrying techniques in modified form from one field to another" and that "we cannot expect answers only from technologies traditionally associated with a problem." [2] While either the rational or the nonrational views of innovation are caricatures when expressed in pure form, my own belief and experience lean toward the second view. This point of view, which seemed to be generally emphasized by our correspondents in this study, will be stressed throughout this paper. It is a view that runs definitely contrary to what I consider to be the current overemphasis on the rational elements of planning and programming of technical work, especially within government.

Research is best regarded as a continuing process involving a series of contingent choices by the researcher. Each time he decides between alternative courses of action, the factors that influence his choice determine the degree to which the research is basic or applied. If each choice is influenced almost entirely by the conceptual structure of the subject rather than by the ultimate utility of the results, then the research is generally said to be basic or fundamental, even though the general subject may relate to possible applications and may be funded with this in mind. The fact that research is basic does not mean that the results lack utility, but only that utility is not the primary factor in the choice of direction for each successive step. The general field in which a scientist chooses or is assigned to work may be influenced by possible or probable applicability, even though the detailed choices of direction may be governed wholly by internal scientific criteria. Research of this type is sometimes referred to as "oriented basic research." Much biomedical re-

[2] Schon, D. F., *op. cit.*, Chap. I.

search is of this character, since almost any new knowledge in the life sciences has a fairly high probability of being applicable.

As another example, once the transistor was discovered, and germanium became technologically important, almost any research on the properties of Group IV semiconducting materials could be considered to be potentially applicable, and this has indeed proved to be the case in practice. On the other hand, research into the theory of zone-refining single crystals was of such obvious immediate application to the control of transistor materials that it could be legitimately called applied rather than merely applicable. Prior to the discovery of the transistor, both of these types of research would have been of equal interest and importance from the scientific viewpoint, but they would have been classified as quite fundamental or "pure." Indeed the same two types of research carried out in a university might be regarded as fairly "pure," while in the Bell Laboratories they would be regarded as "applied" simply because potential customers for the research results existed in the immediate environment. Furthermore, the detailed choices of successive steps in the research would probably be different in the two environments, and might lead ultimately to somewhat different investigations even though they started from the same point. If the next step is toward the particular, the research is more likely to be applied, but if it is toward the general, or toward widening the scope of applicability of a technique or principle, it is more likely to be basic. Thus, in the zone-refining example, the research man in a semiconductor industry might concentrate his attention on study of the purification of promising semiconducting materials, while the university scientist might become interested in exploring a wide variety of materials, both in order to study the dependence of purification efficiency on a wide range of material parameters and to explore a variety of materials of very high purity with a view to determining the sensitivity of various physical properties to purity or crystal perfection.

The essential point is that the categorization of research depends on the existing situation in technology and also on the environment in which it is conducted. As definite categories, basic and applied tend to be meaningless, but as positions on a scale within a given environment they probably do have some significance.

Although basic or fundamental research tends on the average to be less applicable in the sense defined above, the terms basic and applied are, in another sense, not opposites. Work directed toward applied goals can be highly fundamental in character in that it has an important impact on the conceptual structure or outlook of a field. Moreover, the fact that research is of such a nature that it can be applied does not mean that it is not also basic. Almost all Pasteur's work, from the fermentation of beet sugar and the disease of silkworms to the anthrax disease of sheep and the cure of rabies, was on quite practical problems; yet it led to the formulation of new biological principles and the destruction of false ones, which revolutionized the conceptual structure of biology. As another example, studies of semiconductor devices have opened up whole new areas of basic solid-state research that would probably never have been conceived of if the problem or phenomenon hadn't first showed up in a practical device. A good example is the tunnel diode. It was discovered that the current-voltage characteristics of these devices showed peculiar kinks that could be explained in terms of the atomic vibrations of the crystal lattice, and this in turn opened up a whole new technique for the precise study of atomic vibrations in crystals.[8] Similarly, the availability of very pure single-crystal semiconductor materials, which was made economically justifiable only because of their commercial importance, enabled researchers to study the physical and electrical properties of these materials with a precision of detail

[8] Hall, R. N., J. J. Tiemann, H. Ehrenreich, N. Holonyak, Jr., and I. A. Lesk, "Direct Observation of Phonons During Tunneling in Narrow Junction Diodes," *Physical Review Letters*, Vol. 3, No. 4, 1959.

hitherto impossible. This in turn led to deeper understanding of the theory of the behavior of electrons in crystals quite generally, and really opened up an important new area of theoretical research.

Despite this basic-applied feedback in research, if the criterion used is that of the individual choices of the investigator *after* his initial choice of general field of work, then I think a fairly meaningful distinction can be made between basic and applied research.

Industrial researchers are most skeptical of such questions as What proportion of our research should be basic and what proportion applied? or indeed, What should be the proportions between research and development? They would rather argue from the point of view of business objectives: research, development, production, and marketing are part of a continuous process of two-way information flow, and any distinctions that tend to place barriers at particular stages in this process also tend to reduce the effectiveness of all its individual components. On the other hand, if the researcher at the most basic end of the spectrum is continually having to change the direction of his efforts at the behest of market and production needs, his effectiveness is largely destroyed. Thus science, to be effective in the whole process, needs both isolation and communication. The research and development process may be thought of as a long chain, the two ends of which are well separated but nevertheless connected firmly through the intervening length. The man at the application end of the chain must be able to obtain information directly from the scientist, but the feedback along the chain to the scientist must not be so strong as to interfere with the conceptual integrity of what he does.

Some classification of research into basic and applied is probably needed to protect some kinds of research activity from unrealizable expectations. Basic research is that which may take the longest time to come to a utilizable fruition, and must be judged by the scientific criteria of conceptual significance and generality. From applied research one expects a shorter time of payoff but does not necessarily demand generality or high intrinsic scientific interest. The best science is often that which has practical and scientific importance at the same time, but this is relatively rare and cannot be the normal expectation. Much of Irving Langmuir's work had these characteristics. His discovery and working out of the properties of atomic hydrogen, for example, was of high scientific importance, but also led to the development of a radical new method of welding. His work on surface films opened up a whole new area of fundamental science, whose existence was barely suspected, and earned him a Nobel Prize, but also found immediate applications.

Weisskopf has drawn an interesting distinction between "intensive" and "extensive" research.[4] By intensive research he means research aimed at discovering new fundamental laws and formulating new theories of nature. It is usually characterized by very intensive study of a few simplified systems, chosen because they are believed to exhibit the laws or principles of interest in their most generic or easily isolatable form. High-energy physics and modern molecular biology are both examples of areas of intensive research, in which scientists seek to ask and answer a very small number of very fundamental questions. On the other hand, extensive research usually deals with a larger number of questions, which are less fundamental. It aims at elucidating the applicability of fairly well-understood principles and theories to an increasing variety of systems, often of increasing complexity. Extensive research has the characteristic that when a new experimental discovery is made its theoretical explanation is usually found very quickly. Most of chemistry, solid-state physics, and systematic biology are examples of extensive research areas, in which the fundamental principles are

[4] Weisskopf, V. F., in *The Nature of Matter*, L. C. L. Yuan, ed., BNL 888 (T-306) (1965), pp. 24–27.

understood and the task of research is to discover precisely how they apply to real objects or systems. Extensive research is more likely to be related to applications than is intensive research, and it is no accident, therefore, that most of even the most basic industrial research is of the extensive variety. Indeed, in extensive research the possible ramifications of the underlying principles are usually so diverse and varied that considerations of possible applicability are almost necessary to assist in the selection of problems and research directions. The intensive study of the properties of the Group IV and Group III–V semiconductors in the last ten years was not motivated solely by intellectual curiosity, but was also related to their potential or actual technological interest, as well as to their availability.

On the other hand, the distinction between extensive and intensive research is not absolute. Extensive research can be the basis of great new generalizations or theories. A classic example is Darwin's development of the theory of the origin of species by natural selection, which took place through the accumulation of hundreds of detailed observations representing both applications and tests of the new theory.

Although scientists like to emphasize that fundamental research is "free," it is actually, in another sense, a highly disciplined activity. The discipline is provided by the scientific community to which the researcher is related. His choice of problem and direction is heavily conditioned by the social sanctions of this community, the requirements of originality, and scrupulous reference to related and contributing work of others. The scientist takes these external constraints so much for granted that he does not consciously view them as constraints, but his description of his own activities as "free" may be quite misleading to the layman, who takes the description unquestioningly. In applied research the individual is subject to somewhat different constraints, but not necessarily more severe. They are a variable mixture of constraints arising out of science and constraints arising from the institutional environment in which the research is done. Although scientists are strongly self-motivated, they are also sensitive to their audience. The audience of the academic scientist is the worldwide community of his professional colleagues or peers in his own specialty, communicating through the official scientific literature, through scientific meetings, through "invisible colleges" of preprint circulation and correspondence, and through personal contact. To the scientist in a mission-oriented organization, his audience is mixed. It consists partly of his professional community, but also to a great extent of the colleagues and superiors within his own organization. To feel successful he must feel appreciated, and this appreciation is never accorded or felt purely in terms of his contributions to science, no matter how excellent they may be. He responds to subtle clues in his organizational environment that indicate his value to the organization. Thus one finds examples of good scientists who have left industry, not because they didn't have complete freedom to follow their own curiosity, but because they felt their work was not coupled to the organization, regardless of how much it might be recognized outside. Conversely, applied scientists in an academic environment often feel unappreciated, not because of any explicit pressure, but simply because of the general atmosphere, and because of the lack of means for dissemination or utilization of results of their work.

2. RESEARCH ON RESEARCH

There is increasing concern with the need for better understanding of the research process itself. Several of our correspondents have deplored the lack of systematic scholarship on the research and development process — of research about research. Recently there has been an upsurge of interest in this area, but there is still an absence of solid generalizations based on reliable empirical studies. Much knowledge of the research process comes either from the observations of social scientists with minimal knowledge of the substance of the research area they are investigating, or from the anecdotal evidence of scientists and technologists having little appreciation of the standards of historical evidence and often inadequate appreciation of the economic, social, and cultural factors that influence the rate of adoption and application of research results. There is a need for greater involvement of scientists and technologists themselves in the introspective study of the research process, but subject to the critical scrutiny of social scientists or historians. Many scientists and engineers tend to be unwilling to search for consistent patterns of success in research because they realize the importance of fortuitous interconnections and intellectual spontaneity, and they worry lest dissection of the research process squeeze out this spontaneous element and destroy the environment of successful applied research through premature policy application of untested or overgeneralized findings. The very fact that the natural sciences appear to have a mystique, impenetrable to the uninitiated, often tends to generate an unconscious resentment in students of the scientific process who are not themselves scientists. This creates hazards for the management and support of science, both basic and applied, which increase as the total effort grows larger and more visible.

More broadly, we do not know how to measure the efficiency of science, either in relation to technology, or even relative to its own internal goals. As I pointed out in my paper in "Basic Research and National Goals," [5] we have a few measures of the output of the research and development process or of its individual components. Considering the funds that the federal

[5] Brooks, H., "Future Needs for the Support of Basic Research," in *Basic Research and National Goals*, a report to the Committee on Science and Astronautics, U.S. House of Representatives, by the National Academy of Sciences (March 1965), pp. 77–110.

government devotes to such activities, a greater effort should be devoted to objective empirical studies of the process itself. Much of what is available is based on the personal experience of researchers, and is largely anecdotal. Social scientists are well aware that generalizations based on unevaluated, subjective experience can be very misleading, and even wrong. The Defense Department's Project Hindsight[6] is an example of an honest effort to study the output of research and development objectively. Unfortunately, some of the publicity that accompanied the publication of the summary report justifies the apprehensions of scientists regarding the premature and illegitimate deduction of policy implications from such studies. Although conceived originally as a study of management decisions in the research and development process, Project Hindsight was interpreted by some as a general indictment of the value of "undirected" research. Moreover, there was a failure to distinguish between "direction" from above and "motivation" by the broadly defined goals of the organization. In fact, the evidence from the study suggests that "motivated" research was considerably more successful in producing results than either "directed" or "undirected" research. There are several other recent case history studies, such as those published by the Materials Advisory Board of the National Research Council.[7] It is important that the quality of effort in the "science of science" be raised. Too much current work depends on a misplaced emphasis on statistical approaches such as counting of papers or of technological "events" without reference to quality or significance.

There may be value in having case studies that are conducted

[6] Sherwin, C. W., and R. S. Isenson, "Project Hindsight, A Defense Department Study of the Utility of Research," *Science*, Vol. 156, No. 3782, pp. 1571–1577 (1967).
[7] Materials Advisory Board, "Report of the *Ad Hoc* Committee on Principles of Research-Engineering Interaction," ARPA, MAB–222–M, National Academy of Sciences — National Research Council (1966).

entirely internally within large research organizations as a form of self-evaluation or "technical audit." The lack of need to present a good "front" to a sponsoring agency or to higher management may encourage an intellectually more honest approach, and also permit the testing of more daring or tentative hypotheses or management innovations without becoming committed to them in a more public way.

It is important that some case histories, originally prepared by scientists or technologists themselves, be studied and evaluated by trained historians. The case for the utility of research is usually made on the basis of history, especially in the case of basic research. This is really the only solid ground we have, since basic research in general precedes its applications by ten years or more.

However, it is important to bear in mind that history may be an inadequate guide, since the boundaries between science and technology are becoming increasingly blurred. The decreased interval between scientific discovery and widespread application in recent years has been well documented. Furthermore, a number of social factors are progressively altering the nature of the whole technical enterprise — the growth in numbers of technically trained people as a fraction of the work force, particularly in management positions; the growth of higher education, especially the relative growth of graduate and postdoctoral training; the apparent increasing pace of adaptation of social and political institutions to technical change, at least on a small scale; the institutionalization of research and development as an economic activity; the appearance of a scientific equipment industry. In short, there appears to be a strong positive feedback inherent in the growth of science that increases the receptivity of society to the application of scientific findings and methods in almost every aspect of life.[8] Historical studies have generally

[8] Hirschman, A. O., "The Principle of the Hiding Hand," in *The Public Interest*, No. 6 (Winter 1967), pp. 10–23.

pointed up the fact that the development of technology has been surprisingly independent of the development of science, at least in detail. Yet most of the studies on which this conclusion is based come from the nineteenth and early twentieth centuries, and there is evidence that the detailed interconnection of science and technology is becoming much closer, so that many of the most scholarly and solidly based historical studies may have the least relevance to the contemporary scene.

Industries such as communications, electronics, chemicals, pharmaceuticals, computers, and instruments are clearly science-based, and represent the fastest growing parts of our economy. Yet many scholarly studies of innovation are based on experience in the railroad industry, the metallurgical industries, the auto industry, or the farm equipment industry. In many of these industries the underlying technology tends to be "observational" rather than "experimental," in the sense that it is difficult to do meaningful or relevant experiments except at nearly full scale with models of the final product. J. Goldman and D. Frey call attention to the way in which this circumstance has retarded the application of science to manufacturing processes, owing to the high cost of meaningful experimentation.[9]

The relative role of science and technology in the early history of the Industrial Revolution is well expressed in the following quotation from the German engineer Ferdinand Redtenbacher in about 1850:

> The manifold mechanical movements needed for the arrangement of machinery need not always be invented anew. . . . A very exact and complete knowledge of mechanisms already invented is therefore most important in the arrangement of machines. Scientific knowledge is actually of little help, for complex mechanisms are

[9] Goldman, J. E. and D. N. Frey, "Applied Science and Manufacturing Technology," in "Applied Science and Technological Progress," a Report to the Committee on Science and Astronautics, U.S. House of Representatives, by the National Academy of Sciences, USGPO (1967), pp. 255–296.

> evolved not through general powers of thought but by quite special powers of understanding of form, of disposition and of assembly of parts. Whoever is gifted with these powers and has developed them by varied practice will therefore be able to produce many and very ingenious inventions even though almost totally lacking in previous intellectual education; while he who lacks these powers, even though he have other most remarkable diverse gifts, will not yet be in a position to devise even the most insignificant mechanism.

(From F. Redtenbacher, Prizien der Mechanik und des Machinenbaues, Mannheim, 1852).[10]

The general approach to industrial innovation described in the foregoing paragraph is applicable to much of the period of rapid industrial growth in the United States in the nineteenth and early twentieth centuries. It is not without importance today, but it is no longer the central style of innovation. The dominance of this outlook and style in the nineteenth century is illustrated by the fact that even Josiah Willard Gibbs, the greatest theoretical scientist produced by the United States prior to the twentieth century, was awarded his Ph.D. from Yale in 1863 for a thesis on the design of gear trains, a thesis that relied heavily on geometrical visualization of the type described by Redtenbacher.[11]

It seems clear today, however, that a new pattern is emerging in which the "general powers of thought" are replacing the "special powers of understanding of form" as primary generators of industrial innovation. This seems to happen less by a general uniform evolution than by the appearance and rapid growth of new industries with a new style of thought, beginning with the German chemical industry in the late nineteenth century and culminating in the modern computer, electronics, and communications industries. These industries were the first to develop a science base because their underlying technologies could be treated on a laboratory scale.

[10] Quoted in Friedrich Klemm, A *History of Western Technology* (M.I.T. Press, Cambridge, Mass., 1964), p. 318.

[11] Wheeler, Lynde Phelps, *Josiah Willard Gibbs: The History of a Great Mind* (Yale University Press, New Haven, 1951), pp. 32–36.

Studies by the *Scientific American*[12] show that there is a very high correlation between the rate of growth of an industry and its investment in science and technology. This does not necessarily mean that the research investment is the *cause* of growth; the reverse could well be true. But, as this difference in growth rate continues, and as new science-based industries nucleate and develop almost explosively, it seems clear that research-intensive industry will become an increasingly important segment of our economy. And further, these dynamic industries have a tendency to invade the older industries, as illustrated by the invasion of the textile industry by synthetic fibers produced by the chemical industry, or the invasion of electronics and computers into the machine-tool industry and, more recently, into publishing and educational supplies.

The point might also be made that, as technology becomes more sophisticated, it is created to an increasing degree by highly trained people who have a strong bias toward the abstract and the scientific. These people are increasingly penetrating all levels of management, and it seems likely that their viewpoint concerning the relation of science and technology will itself determine the future of this relationship, regardless of what the experience of an earlier era may have indicated about its nature. Each generation has its characteristic intellectual style, and in our own time abstract thought is quite clearly the dominant mode. Within the universities this is the mode that attracts the brightest students and the best minds, and there is evidence that the students are considerably in advance of the faculty in their adoption of this style. In the nineteenth century, Gibbs, the theoretical chemist, began his career as a mechanical inventor because that was the dominant style of his day, and even his most theoretical and abstract work shows the influence of this geometrical and mechanical style of thought. In the mid-twentieth century the intellectual style has been set by physics,

[12] Brochure of *The Scientific American* (1966), "The New Industrial Management."

especially theoretical physics, and there is evidence that this style is beginning to shift toward abstract mathematics. J. von Neumann, one of the great innovators of this century in both pure and applied science, shifted from chemical engineering to pure mathematics early in his career, and was internationally known as a pure mathematician before he turned his hand to technology. Norbert Wiener and Claude Shannon, also among the prime intellectual sources of modern engineering, were first-class pure mathematicians, although they had other inclinations as well. Computers already represent a technology dominated by mathematicians, and throughout many activities of industry and government we see technology increasingly concerned with "software" rather than "hardware," that is, by information processing and the manipulation of symbols rather than by the processing of materials and energy.

3. ROLE OF GOVERNMENT, INDUSTRY, AND UNIVERSITIES IN APPLIED RESEARCH

A number of correspondents in this investigation felt that there was a serious need in public policy for a better delineation of the relative roles of the federal government and industry in the support and performance of applied research. The government was acknowledged to have a responsibility to support fundamental science, especially where it was connected closely with higher education. The government also has a responsibility to support science that directly contributes to public purposes, such as defense, public health, weather forecasting, or environmental improvement. The responsibility of government in the field of primary food production — that is, agricultural and fisheries research — is also universally acknowledged. There was a feeling that the government should not support research in areas in which private industry was active or could be induced to be active through suitable devices of public policy, such as tax incentives or the creation of new markets through purchase of products or services by public authorities. This feeling is based on more than a political bias in favor of free enterprise; it has a solid basis in the nature of the research and development process. Applied research is most effective when it is coupled to a "market" that provides an automatic measure of effectiveness of the end product of research. The existence of a market gives a continuous incentive for self-appraisal, which is often lacking for activities performed in the public sector. When the government supports applied research in an environment that is not organizationally coupled to an end use, it is likely to stray from the mark, and this becomes more of a hazard the closer the research is to application. It is probably no accident that, by and large, government-supported research has been most successful in defense, where the government itself is the customer for the end product. An exception to this general statement is agriculture, where a slow evolution has resulted in extremely effective coupling between public research and private development, production, and marketing. Nevertheless, it is important to note that the government role in agriculture extends well beyond the research itself to include extension services, marketing, economic services, and agricultural subsidies. The latter have had the effect of guaranteeing markets and thus to a considerable extent underwriting the economic risks of innovation. Another good illustration of a desirable pattern is provided by nuclear power. When research and development within the Atomic Energy Commission laboratories reached the stage at which successful development of civilian nuclear power plants seemed likely, the Atomic Energy Act was modified to encourage transfer of the new technology to industry as rapidly as possible, and the criterion of success became the willingness of public utilities to purchase nuclear power plants following their own evaluation of the comparative economics of conventional and nuclear plants. Basically, the criterion for transfer was the willingness of private industry to take on the task, again really a market criterion.

Nuclear power was a development of such magnitude that federal investment in the early stages of the technology was the only way it could be demonstrated. To put the matter in another way, the risks of the development were so high that only the nation as a whole could afford to bear them. The task of government was really to reduce the technical risks to the point where they could realistically be borne by smaller social units. But direct subsidy of research and development is only one of several methods available for spreading risks. Others include tax incentives, allowance for the cost of independent research in the overhead on federal procurement, guaranteeing to the seller a minimum size of market for an innovative product, government underwriting of price stability (a very significant factor in agricultural innovation in the United States), guarantee by the government of minimum performance standards to the purchaser (for example, in connection with new types of municipal waste disposal). The principle involved in each of these examples is that of maximum decentralization of the decision process, the customer having a strong incentive to make a sound and objective evaluation of the economics. A striking example of technical innovation stimulated almost exclusively by subsidizing a market rather than by direct government support of technology is the case of uranium mining. With little more than a guaranteed price and market, the government changed the known supply from extreme scarcity to abundance in about ten years. In fact, this alteration in uranium reserves — largely the result of private exploration, with some help from general geological knowledge developed by government — was as important a factor in the commercial success of nuclear power as was government-supported research and development. This kind of decentralization is also desirable on a smaller scale. For example, spin-off of small science-based firms from government laboratories and larger industry occurs when technology reaches a stage at which the risks of further innovation can be borne by a smaller organizational unit than the parent organization. The process of spin-off is often assisted by the fact that the new entrepreneur may be partially supported by the parent research organization.[13] A spin-off organization serving a specialized sector of the market is often more effective in the later or more applied stages of innovation than is a large organization with much greater technical resources. Only in this way can one explain the success of firms such as Itek, Polaroid, Xerox, or Texas Instruments.

The federal government should adopt a more hospitable attitude toward spin-off of new industry from federally supported technology, including its own laboratories. There is still a widespread belief that ideas resulting from work done at taxpayer expense should be put in the public domain. However, this belief overlooks the fact that the innovator who develops an invention into a commercial product or process and tries it in the marketplace contributes as much or more to technological innovation and economic growth than the originator of the idea.

There remains the question of the role of universities in applied research. Historically the universities have been the major centers of applied research in both agriculture and medicine, although in both these cases a large corollary development activity has grown up in industry. The university research activity has been well coupled to the operational use of the results. In the case of agriculture, this has occurred through the experiment stations and through the extension service, which have made it possible to demonstrate the economic value of the research results rather directly. In medicine the demonstration activity has occurred through the affiliated teaching hospitals.

[13] Brooks, H., "National Science Policy and Technology Transfer," in *Proceedings of a Conference on Technology Transfer and Innovation*, National Science Foundation report NSF 67–5 (1966), pp. 53–63.

Thus one may generalize by saying that a fairly effective system of technology transfer has grown up in the life sciences, which has made it possible to couple applied research in universities to the ultimate user. Although some universities have developed engineering experiment stations, there is not for the most part a strong tradition of applied research in the physical sciences corresponding to that in the life sciences. This results largely from intrinsic differences between the applied life sciences and engineering. Since living systems always exist in many nearly identical exemplifications, a discovery or invention in the life sciences, even when highly specific and applied, also has a high degree of generalizability. A new technique for a surgical operation can be applied immediately in many nearly identical circumstances. A new variety of seed or a new method of cultivation can be disseminated rather readily, and there is often not a large problem of scale-up from the laboratory to operational use. Where there is such scale-up, as in the case of fertilizers, pesticides, drugs, medical instrumentation, or farm machinery, the corresponding development work has been most effectively done by industry. Thus we see that applied research in the academic environment is most effectively done when it is readily generalizable and where problems of scale-up or large-scale production are not of major importance. Such problems usually involve careful timing, scheduling, or programming of research, which tend to be incompatible with the other requirements of the academic environment.

The problem of scale-up involves more than physical size, however. Of equal importance is the problem of scale-up in complexity or "intellectual size." The development of complex systems involves the coordination of many component pieces of a problem and many individual specialities. Often it involves highly sophisticated science or mathematics side by side with rather conventional or mundane design or repetitive analysis. Such a coordinated effort tends to be incompatible with the university environment, with its high turnover of people, its treatment of research as a part-time activity, with the high value it places on individual as opposed to team performance, and with the value it attaches to the proposing of new ideas as compared with critical evaluation and comparison of ideas and their execution in all the most mundane detail. In the future we may expect more enterprises in the life sciences to partake of the same complexity that is now characteristic of many engineering systems. Thus the increasing significance of "intellectual size" in these areas may generate greater reliance on mission-oriented institutions only loosely associated with universities, or completely separate.

When engineering close to production has been done in universities, it is usually in separately organized and staffed contract research centers having a quasi-industrial character. The close association of such centers with universities or technical institutes does assist in recruitment and also provides a source of valuable applied experience for faculty and graduate students, though often to a relatively small fraction and on a somewhat limited basis. The operation of both contract research centers and engineering experiment stations or institutes has been attacked as competing unfairly with private enterprise, and recently there has been a strong trend of opinion both inside and outside universities against the operation of contract centers for applied research by educational institutions. The responsibility for staffing and administering such centers throws a load on already overburdened university administrations and diverts them from tasks more central to their educational and research missions. It often involves the university in direct competition with industry for contracts, and in making evaluative judgments on subcontract performance by industry. If the research is under security classification, or involves dealing with proprietary information, it departs from the academic tradition that all scientific activities that are proper to a university should be open to the free and searching criticism of the entire world scientific community. There is an often justified fear among university faculties that security classification will be used to cover mediocre, routine, or pedestrian work.

When applied research in universities has led to useful new technologies, it has often been that the research was undertaken to serve a purpose internal to the university, or where the application was a direct extension of basic research. Early computer development was carried out in several universities largely for the purpose of providing a better tool for scientific computation in basic research. The nuclear resonance spectrometer, the atomic clock, the maser, and the laser were all logical extensions of basic research already under way. The high-power klystron was developed for particle accelerators for nuclear research. Some fundamental technological developments, particularly in materials and in chemistry, have come from applied university work. Here again, this has usually happened in areas in which the problems of scale-up from the laboratory were minimal. The universities continue to be major sources of innovation in computers, especially on the software side, though the center of gravity has probably shifted to industry.

In general, I believe that more applied research in universities is desirable, when it is appropriate. One might state a general principle as follows: When basic research is to be supported, the burden of proof should lie with those who wish to place it outside the university; when applied research is to be supported, the burden of proof should lie with those who wish to place it inside the university. The following criteria favor university performance of applied research:

1. The results are readily generalizable, as in medical research.
2. The research lends itself to involvement of students, that is, it is not programmed or scheduled to meet deadlines.
3. It is unclassified and not subject to publication restrictions, and is thus open to full scrutiny by scientific peers everywhere.

4. It is a logical extension or outgrowth of basic research under way or already performed in the university.
5. It is of primary benefit to the public sector, or relates to areas of public responsibility.
6. The inventor is on a university faculty.
7. It relates to the development of a fundamentally new technological capability, involving new principles, and of benefit to more than one company or industry.

It is usually desirable that applied work begun in universities should be transferred to industry, where appropriate, as soon as possible, and certainly prior to manufacturing or operations.

On the other hand, in considering what type of institution is appropriate for what type of applied research, an overriding consideration may be the source of the original idea. Experience indicates that an idea seldom thrives if taken out of the hands of its inventor at too early a stage, and invention does not always follow organization charts or formal definitions of mission.

4. DISTINCTION BETWEEN SCIENCE AND TECHNOLOGY

In this area there is a wide variety of views ranging from the opinion that science and technology are really quite separate to the opinion that technology is essentially applied science. The truth is probably intermediate. Part of the variety of views may be due to variations with time. As I have suggested in section 2, much of the technology in the nineteenth century owed little to contemporary science. On the other hand, an increasing component of today's technology is closely dependent on science — if not on contemporary science, then on science ten, twenty, or thirty years old — but theoretical science nevertheless.

Even for the nineteenth century it is easy to exaggerate the independence of science and technology. Although this tended to be true of mechanical invention, it was less true of the applications of electricity and chemistry, even then. For example, a working model of an electric generator was constructed only a year after Faraday's discovery of electromagnetic induction, and an electric motor two years later.[14] That Faraday's discovery did not immediately turn into a major industry was not due to failure to realize its technological potential, but rather to the fact that a whole complex of other inventions for the utilization of electricity would be required before an economically viable technology could be created.

It is true that there is inevitably a considerable component of "art" in technology. Technology is essentially a codified way of doing things, and much of this is based on systematic theoretical knowledge, which is science, but some simply on codified experience, which is what I mean by "art." A good technologist must sometimes be willing to accept or search for solutions that work, even if they are not fully understood. In this he is not so far from the experimental pure scientist who often behaves like a technologist with respect to his own experimental techniques. In fact, each branch of science is based on a characteristic technology, which changes as the science advances. On the other hand, the greatest impact of the scientist in an industrial environment has resulted from his unwillingness to accept rules of thumb or procedures that are not understood.

The technology associated with an experimental science tends to be passed from worker to worker somewhat independently of the conceptual scheme of the science. There is a collection of "tricks of the trade," which lie outside the body of formal scientific literature. Technologies developed for scientific purposes often later grow into technologies useful for industrial or other operational purposes. Research instruments are first commercialized, then used in other sciences, and finally used to control production processes. Laboratory tools and techniques such as high pressures, cryogenics, high vacuum, spectroscopy, vapor-phase chromatography, and so on, begin in a research laboratory but often end up on the production line. One of the most dramatic examples is the cathode-ray tube which, originating as

[14] See reference 10, p. 348.

a physics laboratory device, became the basis of the modern television picture tube. These experimental technologies undergo transformation and improvement in the process of being applied, but their origin in experimental pure science is still evident.

It also happens, of course, that technologies developed for applied purposes are later turned to providing new instrumentation for pure science. World War II produced a host of new techniques, especially in connection with microwaves, that have become indispensable laboratory tools of physics research, and more recently of physical chemistry. Within the last ten years some of the tools of pure science have become major engineering projects in their own right. The most dramatic examples are high-energy accelerators, satellites for instrumented space exploration, modern radio-telescopes, and the Mohole project (until its tragic demise). In addition, government-supported pure research has created a large commercial market for research instrumentation, including moderate-size accelerators for low-energy nuclear physics.

There are cases in which it may be desirable to develop a field of pure science partly for the sake of the by-product technology that it generates. Although, in principle, this technology might be developed for its own sake without the associated science, in practice the scientific end use provides the focus and motivation, which generalized development could not do. In addition, it attracts more dedicated and able people through the intellectual challenge of the science. Often the other uses of the tools so developed do not become apparent until after the development has been completed. A dramatic example is the oceanographic research submarine *Alvin.* Designed with the needs of basic oceanography in mind, it was used with its research crew to locate the lost hydrogen weapon off the coast of Spain in 1965. Today, in elementary-particle physics, the requirements for information-handling of the tremendous volume of data obtained are stretching computer technology in a way that is advancing the whole art. Already techniques of "pattern recognition" originally developed for automatic scanning of cloud-chamber photographs of nuclear-particle tracks are finding application in other areas, such as automatic letter sorting. Nuclear physics and, more recently, radio astronomy are remarkable for the complexity and volume of the systems of data with which they have to cope. On the other hand, it would be dangerous and misleading to argue for support of these fields solely or even principally because of their demands on data processing. Nevertheless, these by-product effects should properly be considered in assessing priorities or in assessing the relevance of a field to applications. Another example may lie in the field of sensitive optical or infrared detection for the purposes of astronomy. While one would not undertake such technological developments unless they served a very important scientific purpose, their other uses may be of great significance, and the cumulative effects of such technological developments emerging as a by-product of support of pure science over a long period may be very large.

Numerous observers have commented upon the differences between the communications systems within science and those within technology. Science has an elaborate system of public documentation with strong sanctions operating on the individual scientist to make full use of and give proper credit for previous work relevant to his own. Discovery and innovation within science have extremely rapid diffusion times, and the rate of diffusion is influenced to only a minor degree by political and organizational boundaries. There are somewhat higher boundaries between scientific disciplines, but even this is often exaggerated by outside observers.[15] Within technology the communications pattern tends to be more localized and more confined to organizational channels. One reason for this, of course, is that much technological innovation is harder to verbalize and

[15] Oettinger, A. G., "Essay in Information Retrieval or Birth of a Myth," *Information and Control,* Vol. 8 (1965), p. 64.

to document. The intuitive aspect of invention, so eloquently described in my earlier quotation from Redtenbacher, makes it more dependent on face-to-face contact and learning by doing. Thus, typically, in an industrial or government laboratory the scientists tend to be oriented toward external communication and toward recognition and appreciation by the outside professional community, while the technologists tend to be oriented toward internal communications and toward recognition and reward within the organization. I believe a changing pattern is discernible in this respect, however. The newer technologies such as nuclear energy, electronics, and computers tend to be more externally oriented than the older technologies. There is more conscious effort to conceptualize technological knowledge. Although invention in the sense of intuitive synthesis may be just as important in the newer technologies as in the mechanical technology of the past, there is a greater tendency to document and to generalize specific advances — for example, to convert the conception of a particular device into a design theory or into a set of formalized design principles for a class of devices or processes. In my opinion, this is an important trend, and one that is highly desirable from the standpoint of accelerating the application of science and the diffusion and adoption of technical innovations.

In the technology of medical care, the tradition of documentation of practical advances is somewhat better established than in engineering and industrial technology. This is partly because, as mentioned earlier, medicine is dealing with the same human body everywhere, so that medical procedures may be more generalizable than engineering procedures, which relate to the design of particular products or processes.

A particular problem in the interaction between science and technology has been eloquently described by Peter Drucker.[16]

[16] Drucker, P. F., "The Technological Revolution: Notes on the Relationship of Technology, Science, and Culture," *Technology and Culture*, Vol. 2 (1961), 342–351.

It is the reluctance of technologists to deal with the more mundane and less sophisticated problems, which still may be quite important socially. This is a special difficulty in connection with the transfer or adaptation of technology to underdeveloped countries, but it is also an inhibition against application of technology to the more backward civilian industries in our own country. The problem is probably closely related to the discussion by Cyril Smith[17] of the reluctance of scientists to deal with complex or "messy" systems. The inhibitions are undoubtedly associated with the fact that the solutions, even when successful, are seldom elegant or intellectually satisfying. The importance of such problems is not in itself sufficient motivation for attention when there are comparably important problems in more sophisticated areas that give greater intellectual satisfaction.

5. APPLIED RESEARCH AND BASIC ISSUES

As applied research and development are more and more performed by people with original training in basic science, and thus interested in and aware of basic issues, applied research is likely to bring increasing benefits to science itself, as well as to technology. Applied research will continue to turn up important basic issues that the discoverers will increasingly be capable of recognizing and pursuing. This will be recognized also as having benefits for technology itself, for when applied problems are approached with the methods and the generalizing tendencies of basic research, the solutions found tend to be more broadly applicable, or to lead, by "serendipity," to new applications. Applied research must often look beyond the time horizon of the immediate purpose for which it is undertaken. The more sophisticated the field of application the less likely it is that the first version of a new invention will be valuable without much

[17] C. S. Smith, in "Applied Science and Technological Progress," a report to the Committee on Science and Astronautics, U.S. House of Representatives, by the National Academy of Sciences (June 1967), pp. 57–71.

further development. It is in this further development that applied research aimed at deeper understanding of the underlying phenomena is especially important. For example, the first discovery of the gas laser at Bell Laboratories was followed by an intensive period of rather fundamental research in atomic and molecular physics, which eventually led to greatly improved lasers, culminating in the development of the CO_2 laser with a power output several orders of magnitude greater than that of the earliest gas lasers.[18]

A fundamental problem in the education of the modern applied scientist is how to train him to bring a basic research viewpoint and approach to applied science without creating in him a disdain for, or impatience with, applied problems. A frequent shortcoming of the basic research viewpoint is a tendency to view all problems in the light of the researcher's own specialty. Although basic research training is undoubtedly the best possible type of training for the ablest people, who are able to turn their attention to many different classes of problems, applying the intellectual techniques they have learned through deep study of one specialty, the less capable people may tend to become prisoners of their specialized training, choosing specialization as the best way to exploit their more limited capabilities. In the best of American graduate education this is offset by requirements of a certain degree of breadth in course work, but it is a constant hazard of graduate work — paradoxically enough, even more often in engineering than in the sciences.

Enlightened industrial laboratories often adopt the practice of encouraging newly hired Ph.D.'s to tackle problems quite remote from the areas of their thesis research. The value of graduate training should lie partly in the confidence it instills in the student to solve new and challenging problems, and to assemble independently the information and tools necessary to

[18] Cf. article by W. O. Baker, "Science and National Security," in "Research in the Service of National Purpose," ed. by F. J. Weyl, Office of Naval Research, Washington (1966), pp. 117–118.

do it; yet too many students want to use their first work assignment as an opportunity to extend and improve upon their Ph.D. theses, rather than to broaden their experience and skills.

6. INVOLVEMENT OF NONSCIENTISTS IN TECHNOLOGICAL JUDGMENTS

When the experts disagree on the correct technical course to take, the decision between alternatives is often thrown back on the legislator or nonscientific executive. The question of the proper degree of involvement of the nonexpert in technical judgment is one of continuing controversy. As with all arts, executives and legislators with long experience develop a surprising talent for ferreting out key technical issues, without understanding the technicalities. The congressional hearings on the nuclear-test-ban treaty provide a classic example of how it is possible for the experienced generalist to elucidate the key issues without necessarily understanding all the underlying science. Similarly, examiners in the Bureau of the Budget, though trained as political scientists or economists, acquire an uncanny "nose" for the important issues without really being able to argue the technical merits.

Furthermore, many of the types of questions that legislators or executives are required to answer are really questions of political preference, which are only slightly disguised as technical issues. Most commonly, important decisions in applied science depend not on technical feasibility, which is uniquely the province of scientists and technologists, but on social desirability, which must be determined by a multidimensional interaction of scientists, technologists, public servants, and the public. In practice, questions of technical feasibility and cost interact with desirability, and hence the need for a many-sided discussion. A good example of the type of complexity involved is the case of the supersonic transport. Real dangers are involved, however, when the nonscientist attempts to impose his own value system on what should be largely scientific decisions. The public is often tempted to dump large amounts of money into the solution of problems that are perceived to be of social importance, without adequate consideration of feasibility or economic efficiency, and without adequate understanding of the interrelationships within science. The result is sometimes an assignment of relative priorities, which actually diverts resources from what might be the most promising lines of advance. The national investment in aircraft nuclear propulsion is probably one of the most striking examples of such misapplied effort. There is a special hazard of misconceived priorities in the field of health, in which the most "visible" diseases tend to receive the greatest research attention. In times of rapidly increasing budgets this is not a major problem, because there is usually enough surplus to cover the less spectacular but scientifically more promising efforts. Furthermore, individual scientists have their own sense of priorities, and the most creative people tend to allocate their effort to the most productive fields as long as some support is available. The danger comes when funds become so tight that narrowly conceived applied efforts completely pre-empt the field.

In the past fifteen years the legislator and nonscientific administrator have tended to adopt a "hands off" attitude toward scientific decisions, probably more than is desirable even in the long-range interest of science itself, but currently the pendulum seems to be swinging rather violently in the opposite direction. The nonscientist often tends to see utility in rather narrow terms, and can be impatient with what appear to him to be diversions from the principal social goals in the name of science. There is a sort of corruption here which is dangerous for science and technology. The availability of large resoures for efforts of apparent social importance may tempt scientists to make expedient promises of quick utility in order to obtain support for work they wish to undertake. It is always relatively easy to invent new terminology to label fundamental scientific work with perfectly legitimate "applied-sounding" words. But the unfortunate aspect of this is that, usually, the less the ability and integrity of the scientist the more willing he is to invent expedient labels for his work, so that the net effect of providing support preferentially for fields or projects that have the appearance of immediate social utility is to drive the best and most creative minds out of the field. In many instances it may be better to support fully a first-class scientist who is willing to devote part of his effort and thought to the applied needs of an agency than to support several mediocre scientists who are willing to devote their full effort to the problem as defined by the agency. Intellectual freedom is more valuable than money in attracting first-class people into socially important fields, and Congress would do well to keep this in mind when providing funds for applied work. The more narrowly the objectives of a program are defined, the more likely it is to drive out the most creative workers. Even for applied work, the system of research evaluation and project selection by judgment of peers is a powerful antidote to the waste of money on work that is spuriously claimed to have immediate social utility.

On the other hand, I would agree with those who say that scientists and engineers have a much greater obligation than they have assumed in the past to explain their work in terms that are intelligible to the nonexpert and the general public, without being condescending. Too many scientists confuse simplification with condescension. There is good intellectual discipline in explaining oneself to people not committed to one's own specialty. However, it is essential that short-term support decisions should not depend primarily on *annual* justification to nonscientists. The cycle time for such justifications should usually be several years.

The postwar era was characterized by heavy attention to technical problems related to national security — defense, atomic energy, and space. These problems are technology-limited to such an extent that social and economic considerations are of minor or negligible importance. The characteristic institution of this period was the agency defined by an area of technology, for example, the Atomic Energy Commission or the National Aeronautics and Space Administration. In the last few years, public and political attention have turned toward problems having both technological and social components, usually in complex admixtures: transportation, urban reconstruction, pollution, education, and industrial growth in lagging sectors. Even in the defense area, the shift of emphasis to limited war has brought in a much larger social component; the field of health, the only other area of major federal research investment, partakes of some of the characteristics of defense, being largely science-limited at the present stage. In health, however, emphasis is shifting from the understanding and cure of disease somewhat toward the organization and delivery of care, again having a larger social component. For such society-limited problems, a factor of considerable importance is the social acceptability of solutions to many people, something which was of little or no concern in the Apollo program or the Minute-Man weapons system.

The distinction between science-limited and society-limited problems is not invariant with time, and may in fact be radically altered by technological progress, as Weinberg[19] emphasizes in his paper. A new technology can overcome social limitations in several ways: by drastically reducing the cost of certain operations or products, by greatly simplifying certain products or operations and thus making them more accessible to the average individual, or by developing a wholly different way of doing things that does not have the same side effects as existing procedures. Classic examples of cost reduction were the invention of barbed wire, which made fencing economically feasible in the

[19] A. Weinberg, in "Applied Science and Technological Progress," ref. 17, pp. 415–434.

American West; mass production, which made personalized transportation available to the average man; and automatic telephone switching, which made electrical communication accessible to practically everyone. An example of simplification is the IUD (intrauterine device) which has greatly reduced the motivation necessary for family planning in underdeveloped countries. Examples of alternate ways of doing things are the introduction of nuclear power, which makes possible energy generation without atmospheric pollution, or the development of biological methods of pest control, such as male sterilization, or, in the future, species-specific hormones, which permit elimination of pests with less serious environmental consequences than do general-purpose chemical pesticides.

Inventions that permit "designing around" social obstacles require just as much social ingenuity as technical ingenuity, and often the two have to be combined in a single individual. The process of inventing a product for a market is usually one that requires both technical and social invention. The perception of a market possibility consists in seeing what kind of technological invention is needed to overcome a particular social barrier. There are times when stating the need for a particular invention without any knowledge of how it can be done technologically may be a much more important step than the technological solution itself. This was very clearly understood in the nineteenth century, when many inventions were of this character and required relatively little sophistication in technology.

It is perhaps a hazard in today's highly sophisticated world that preoccupation with technology — a preoccupation made necessary by the high level of education required — may result in too little recognition of the equally important necessity of properly articulating social needs, or, if you prefer, the requirements of the market. With respect to the great modern problems — what I call the four P's of population, pollution, peace, and poverty — it may be that articulating these is the most important part of the problem: that once these needs are formulated in the right way, the technological solutions will become obvious, or will fall into place. However, it is important to note in this connection that what is required is the articulation of social needs in such a way that we can get from where we are to where we want to be without the necessity for massive persuasion or massive education of many people simultaneously. Technology can lower the "activation barrier" for such a process so that we can get from where we are to where we want to be by running downhill gradually rather than by crossing a mountain pass of social acceptance.

7. THE MISSION-ORIENTED LABORATORY

The characteristic institution for the conduct of applied research in the modern era is the large, multidisciplinary mission-oriented research organization. Although this type of organization has not replaced the small specialty company or even the independent inventor as a source of innovation, it is to an increasing degree the source of basic technology both for public purposes and for industrial projects. The term "mission-oriented laboratory" comprises several different types of institutions:

a. Large federal civil service laboratories, such as the major laboratories of the Department of Defense, the various National Aeronautics and Space Administration centers, and the pioneering laboratories of the Department of Agriculture.
b. Federal contract research centers operated by universities, university research associations, or separately organized nonprofit corporations. Examples include the Lincoln Laboratory of the Massachusetts Institute of Technology, the Los Alamos and Livermore Laboratories of the University of California, the RAND Corporation, the Aerospace Corporation, and the Institute for Defense Analyses.

c. Central industrial laboratories or federal contract industrial laboratories such as the Bell Telephone Laboratories, the General Electric Research and Development Center, the Sandia Corporation, or the Oak Ridge National Laboratory.

d. Multipurpose research institutes or research service industries such as the Stanford Research Institute, the Battelle Memorial Institute, or the A. D. Little Company. These organizations provide research services to clients on a job-shop basis.

I have deliberately not included in this classification research institutes whose principal "mission" is the advancement of a field of pure science rather than the development of technology or the fulfillment of some social mission such as defense, health, or food supply. Examples are the Brookhaven National Laboratory, the Carnegie Institution of Washington, the Smithsonian Institution, the Woods Hole Oceanographic Institution, or the National Radio Astronomy Observatory. Admittedly, the line between these and mission-oriented laboratories is sometimes hard to draw, since the mission-oriented laboratories carry out much fundamental research bearing on their mission, while the scientific institutes often move close to technology or utilize technology in the design and development of their research equipment. An organization such as the Jet Propulsion Laboratory of the California Institute of Technology is especially difficult to characterize, since its operations are heavily technological, but its basic mission essentially scientific.

What constitutes a "mission"? How is it defined, and how is it used to shape the specific research program? How is success in the performance of a mission to be measured?

The answers to these questions are complex and often subtle. A mission must be neither too vague nor too specific. It must be concrete enough to provide real guidance in the choice of tasks and priorities, and to be understandable by the key people in the organization, but it must be general enough to permit the phasing-out of old tasks and the establishment of new research goals. A mission must be like the shell of a building, within which the interior can be drastically rearranged to carry out constantly changing tasks. A mission, however, should not be simply an umbrella under which almost any high-quality scientific activity can be justified. Not every exciting discovery is convertible into an economically or socially useful product. Unfortunately, the broader the objectives of an institution, the harder it is to determine what is really relevant to its mission. Very large diversified companies find that almost everything is relevant in principle, but they have to pick and choose, at least in the short run, in order to achieve "critical size" in the efforts they do support. In many cases it may be more important to maintain this critical size than to "cover every bet." One reason for this is that the transfer of information between organizations occurs more rapidly, except under conditions of secrecy, than does the vertical transfer from research or invention to marketable product. In the research part of an institution, it is sometimes more important that the organization be working in a general field than that it be working on a particular project. A company — or for that matter, a nation — that has a broad technical capability can quickly exploit the ideas of others, and can catch up on the bets that it misses provided it has the technical sophistication to identify promising ideas at a sufficiently early stage. The example of the tunnel diode, discussed in the paper by Suits and Bueche,[20] is a good illustration of how an organization can rapidly convert an external discovery into a valuable product. It happened essentially because there were people inside General Electric who were capable of making the same discovery — indeed, might have made it. From the standpoint of the effectiveness of the organization, it was more important that they were capable of making it than that they actually made it, because this guaranteed that they would recog-

[20] C. G. Suits and A. M. Bueche, in "Applied Science and Technological Progress," ref. 17, pp. 297–346.

nize the significance of the discovery as soon as it appeared. Just as a company or a nation cannot expect to exploit every promising scientific discovery, so every discovery that it exploits need not be its own.

In considering the "missions" of government laboratories, it is essential to distinguish a "mission" from a "task." A mission is a function assigned to an organization by higher authority or by legislation. A task is a subordinate objective that is best generated from within the research organization and pursued, usually by agreement with the sponsoring agency. A research institute that does not generate most of its own tasks, but depends on external direction or "orders from headquarters," is either suffering from inadequate leadership or has a mission which is inadequately defined.

The definition of its mission is one of the most important considerations in establishing a new research organization or reorienting an old one. In evaluating the performance of such an organization in applied research, the emphasis should be on the performance of the organization as a whole rather than on its individual components. Good applied research is of little value if the mechanisms do not exist to translate research results into goods, services, or operations. A frequent paradox observed in civil service laboratories is the high level of scientific performance of individuals contrasted with the often disappointing results from the organization. A good scientific performance often does not add up to an effective overall performance, partly because of the cumbersomeness of the decision-making process, and partly because of poor communication between the working scientist and the headquarters organization that supports and administers his work. Too often the insights and understanding of the working scientist are distorted and transformed by many levels of intermediaries, each perfectly competent and honest in itself. Scientific advice transmitted through many levels of management loses its intellectual integrity, and hence its effectiveness. For similar reasons, the working scientist in government laboratories fails to acquire a comprehensive view of the goals of his laboratory and how his efforts contribute to them. Not fully appreciating or understanding the broad goals, he is unable to identify the key tasks and to establish his own set of priorities consistent with the broader goals. These faults are by no means confined to government laboratories, but the strongly hierarchical nature of government tends to aggravate the problem. Too often in government, initiative and reporting in a scientific organization are associated with position rather than with specific subject-matter competence.

There are certain identifiable characteristics of successful mission-oriented laboratories that seem to be independent of whether they are located in government, industry, or universities. These characteristics are more related to the "sociology" or the communications pattern of the institution than to its formal organization. I have listed some of these characteristics below, with a brief discussion of each:

1. Full awareness and general acceptance of the principal goals of the organization by its key people.

   Even though the motivations of the sponsor and the scientist may differ, the scientist is aware of the main problems and goals of the institution. The goals are arrived at by discussion and argument and are usually more the result of consensus than of directives passed down from the head, although the consensus may be formally ratified by such directives. Successful organizations have a tangible "organizational memory," a sort of collective recollection of enduring themes and problems.

2. Willingness to consider and implement new ideas and initiatives on their own merits, regardless of the organizational level at which they originate, or whether they come from inside or outside the organization.

   The phenomenon of "not invented here" is one of the

worst enemies of innovation, and one of the hardest to overcome. The many case histories discussed in this series of papers show that no part of an organization has a monopoly on new ideas. Furthermore, ideas must be nurtured in their early stages by the group that originates them, regardless of its formal task or mission. The originating group should be permitted to carry a new idea to the point where it is either proved unpromising or sufficiently demonstrated to convince a more appropriate part of the organization to adopt it.

3. Mobility of people between the more fundamental and applied activities of the organization.

   As emphasized frequently by our correspondents, as well as by the papers in this series, people are the best carriers of technology transfer, and therefore the organization should be flexible enough so that people can move with ideas and technologies. Many mission-oriented organizations take on young scientists as basic researchers and then simply permit the environment to attract them into more applied activities as they gain in maturity and experience. Since new technologies frequently derive from laboratory techniques, it is natural that the young basic scientist of today may become the successful applied scientist of ten years from now, provided the environment is such that he feels rewarded by being able to contribute to the applied problems of the organization. Not all basic researchers will move in this way, but a considerable percentage will, and will be replaced by new young scientists.

4. Quick recognition and funding of new ideas, at least to the point of ascertaining the desirability of a larger commitment.

   This is probably the single most important factor in technology transfer. Studies of innovation indicate over and over again that the successful ideas are nearly always those for which initial funding was obtained promptly and with a minimum of review by higher authority. Obviously major commitments have to be reviewed, since no organization can develop more than a small fraction of the promising ideas it originates, but each level in the organization should have sufficient flexibility in funding to be able to commit some fraction of its budget to a new idea on its own responsibility.

5. Extensive freedom at each organizational level in the organization to reallocate the resources within its relevant area of responsibility.

This is, of course, a close corollary to number 4. Each level of management in research should have the largest possible measure of control over all the resources needed for effective performance of its job. Multiple functional restraints in travel, telephone, supplies, personnel ceilings, etc., which are not directly relevant to assigned task responsibilities, should be avoided. In government laboratories such restraints are often imposed by administrators or by legislation as a form of cost discipline that substitutes for the more automatic cost controls that operate in the private sector, subject to the discipline of the market. In practice, however, such controls frequently lead to inefficiencies in research performance, whose cost far exceeds the savings resulting from the controls. The situation in government laboratories has improved greatly in recent years, but much remains to be done to give the laboratory or program manager greater freedom to use his resources under an overall fiscal ceiling. This is not to say a manager should not operate under a functional budget, but this budget should be a guide and not a straitjacket, and each reallocation among functional categories should not be subject to prior review by higher authority. The proper substitute for functional constraints is better definition of substantive goals, and measurement of performance against these goals at each level. Furthermore, if people in a laboratory are to be realistically measured against such goals in their performance, they themselves must have a part in setting the goals, as I have indicated under number 1 above.

6. Full communication through all stages of the research and development process from early research to ultimate user.

   The various stages of the research and development process should overlap substantially. The studies under Project Hindsight showed that many important innovations in weapons systems occurred after the design was fairly well committed. This underlines the importance of continuing research well after designs have been frozen, as a hedge against unforeseen problems or difficulties. Similarly, wherever possible, the ultimate user should be brought into the picture as early as possible. This procedure has been especially well developed in the chemical industry, where the operator of the plant often joins the design team at an early stage. The research and development people should be encouraged to follow products and processes into ultimate use in order to obtain "feedback" on problems and potential improvements.

7. A good organizational memory for the enduring technological problems and themes related to the broad mission of the organization or laboratory.

   It is this organizational memory that most distinguishes a mission-oriented laboratory from a university or basic research laboratory. In basic research, memory and continuity tend to be deposited in the scientific literature and the professional communications system rather than in a particular organization. Suits and Bueche[20] in their paper emphasize the role of this organizational memory in several of their case histories, especially that of the vacuum switch. It is partly what distinguishes the "professionalism" referred to by Bode[21] in his paper. Of course, this memory can also become a dead hand, a knowledge of what can't be done, if it is not combined with a high receptivity to new ideas. Mission-oriented laboratories must thus combine continuity and mobility of personnel. Some people will move with new technologies; others may live out their careers with particular constellations of problems or themes.

Isolated discoveries are of little value unless they fit into a communications pattern. This can be either the pattern of the scientific community or that of the "mission" of the institution. Science is transformed into technology through the intersection of the scientific pattern of communications with the institutional or mission pattern. It is the institutional environment that provides the mechanism for identifying the significant problems — significant, that is, in terms of the mission — and this is where the importance of the organizational memory lies.

8. A system of recognition and reward that assigns highest significance to technical contributions to the goals of the organization.

This is also a corollary of point number 2 above. If the sources of innovation are not associated with position in the organization, then it follows that the system of reward and recognition should not be associated exclusively with organizational responsibility. The individual who makes an outstanding technical contribution, but prefers to continue in the laboratory with a few colleagues, should be as eligible for recognition, financial and otherwise, as the individual who manages a large group. A more subtle question is that of the degree to which institutional recognition should reflect external professional recognition. Certainly, professional reputation must be an important factor, but the mission-oriented laboratory should offer comparable re-

[21] H. W. Bode, in "Applied Science and Technological Progress," ref. 17, pp. 73–94.

wards to the individual who chooses to follow the pattern of the internal or institutional communications system rather than the professional one.

A continuing problem for large laboratories that are not directly coupled to an industry or an operating organization is the age structure of the laboratory. This problem becomes especially acute in the case of government laboratories, whether under contract or in the civil service, because of the relative lack of mobility of personnel. Many such laboratories have grown rapidly in recent years and have a relatively young average age, but now there is a tendency for growth to level off, and with it the intake of younger scientists. As the average age of the laboratory increases, its outlook and style tend to change. It becomes more "professional" in the sense of having a deeper background in its assigned technology and mission, but it also tends to become more conservative and resistant to new ideas. The lack of infusion of young scientists with new techniques and viewpoints acquired from their training lowers the innovative spirit of the organization. When the laboratory is coupled to an industry, a certain proportion of the people move on to other company activities, either more specialized manufacturing support laboratories or operating organizations. Here they infuse a new spirit into company management and operations while leaving room for new blood to come in at the bottom in the research activity. Similarly, in universities there is a constant flow of young people through the lower ranks of the organization, and, although the system of tenure introduces some of the rigidities characteristic of the civil service personnel structure, there is generally greater mobility in universities than in mission-oriented laboratories.

However, aging of the staff is only one of the pitfalls of such laboratories. Their mission may become obsolescent, owing either to technological progress in other areas or to changing external requirements and emphases. Yet these changes can be very subtle. A laboratory can retain much of its original technical competence, while its mission is no longer quite adequate to fully engage its capabilities. Under such circumstances, the most creative people in the organization may tend to leave or to drift more and more toward basic research as a substitute for other mission goals. While basic research is essential to the technical vitality of a mission-oriented laboratory, it should not become an end in itself in that setting.

8. STATUS OF APPLIED SCIENCE IN THE UNITED STATES

After many decades of inadequate attention to basic science, and inadequate recognition for scientists, the United States may have overreacted. Raising the status of applied science in universities has become a real problem, as both our correspondents and our authors have suggested frequently. There has always been a kind of status hierarchy of the sciences, in order of decreasing abstractness and increasing immediacy of applicability. This hierarchy begins with pure mathematics and runs to theoretical physics, experimental physics, chemistry, biology, engineering, medicine, social sciences, and so on. Even within a single discipline, one finds a kind of hierarchy between the pure and more applied practitioners of the discipline. In a sense, what has really happened is a disturbance of an equilibrium that has existed for a long time. Historically a certain snobbery has always existed between pure and applied science, but whereas once the exponents of pure science represented a small and rather ascetic minority who took pride in sacrificing the greater material rewards of applied science for what were thought of as the greater psychological satisfactions of pure science, today the gaps in both material reward and external prestige between basic and applied science have largely disappeared. However, some of the mythology of "we happy few" in basic science has persisted. As pointed out by many observers, the values and attitudes of the academic "subculture" are increasingly becoming the dominant values of the influential segment of society as a whole, perhaps inevitably in a time when more than 50 percent of high school graduates go on to college. The business tycoon of the last generation boasts of his son's achievements as a theoretical physicist or a Sanskrit scholar, and economic motivations no longer play as large a part in affluent middle-class society.

Historically, in the United States, training and recruitment of people for the applied sciences has usually been associated with the progressive democratization of higher education. The most dramatic example is the establishment of the land-grant colleges and universities. But in more recent times it is notable that the G.I. Bill resulted in the recruitment of thousands of able students into engineering and other applied sciences. With the exception of medicine, entry into the more applied aspects of science has always been an important path for upward social mobility in American society, and the applied sciences and engineering have flourished in periods during which political and social forces have brought new groups of people into the professional and technical labor force. Applied science benefited greatly from the influx of people that occurred as a result of the wartime mixing up of the normal status relationships of American society. It may well be that the present drive toward equality of opportunity in education will prove one of the most important and beneficial influences in raising the level of applied science in the United States in the coming years.

One cannot discuss the question of basic versus applied science without recognizing that there is some objective basis for the snobbery that exists among academic people. This does not mean that it is right or justified, but, on the basis of numbers of people alone, it is inevitable that on the average basic research will have a higher percentage of able people than applied research. Since there are fewer basic researchers, the group is more highly selected. Furthermore, basic research on the average probably *is* more demanding than applied research. This is simply because much applied research can be directed and organized to a greater degree than basic research; thus people with high skill in the *techniques* of a discipline, but little creative imagination, can perform very well in applied research, given imaginative and energetic leadership. The catch is that the leadership function in applied research is probably *more* intellectually demanding than that in most basic research. Thus the professor who advises his less able Ph.D. students to go into applied research may be acting from a sound instinct. Under first-class leadership, his less able student may have a more productive career in applied research, especially in comparison with academic research, which tends to be highly individualistic. The professor's advice, while sound from the standpoint of the individual student, may adversely influence the general climate of opinion among students, and ultimately, colleagues.

While I agree as a practical matter that the difference between basic and applied research is somewhat illusory, the snobbery is too persistent and real to be dismissed by definition. It is not only the result of willful ignorance among academics, although this doubtless plays its part. In my view, the basic-applied tension will always exist, and is actually a creative force for innovation. The tension should be exploited, not denounced. The important thing is that academic-industrial and the academic-government interactions should be fostered wherever possible, and every effort should be made also to foster a high mobility of scientists between government, industry, and universities both for short periods and on a long-term basis. My own experience suggests that academic scientists do not shy away from applied work when they see a genuine opportunity to contribute, and when they have "opposite numbers" in industry and government with whom they can communicate effectively. Applied scientists, on their side, have a responsibility to reformulate some of their major problems from a basic viewpoint, so that they appear more intellectually rewarding. It is only in this way that the road from science to technology can be smoothed.

Applied science in the United States is most lagging in the areas of technology that have fallen so far behind the research frontiers of science and engineering that effective communication between the technologist and the scientist can no longer take place. It may well be that these areas of technology can be reformed only by invasion, by people moving in from relatively remotely related disciplines and seeing the problems and opportunities of an old technology in a new light.

9. JUDGING APPLIED RESEARCH

Quality is very much more difficult to assess in applied science than in fundamental science. If one asks a random sample of scientists who the best people are in their fields, there will be a surprisingly high degree of agreement, but it is much harder to achieve such unanimity in applied science or engineering. Indeed, the scientists in a given area of work are much less likely to know each other or each other's work in applied science than in basic science. This is partly because in technology the method of communication is much more by personal contact than by literature. Documentation, especially public documentation, of new ideas, is given much less attention by technologists than by scientists. In tracing the history of innovation, one is struck by the frequency of reinvention of ideas by different groups without knowledge of each other's work. Technology, having a higher component of "art," is much more difficult to codify in formal "literature." Good applied science has to be judged by a mixture of criteria, including criteria internal to science itself but also external to science, in varying proportions, depending somewhat on the closeness of the work to immediate application. In judging applied work, quality has to be defined to a large extent in terms of pertinence to the problem at hand. The further removed one is from problem solving, however, the more important internal scientific criteria become in judging the quality even of applied research.

It is probably important that as much applied research as possible be translated into the "scientific" tradition of public documentation, with as much emphasis as possible on generic solutions. To an extent that is often neglected, this documentation function in the industrial sector is performed by trade journals, which are less formal than the "official" scientific literature but perform an essential function in diffusing new technology throughout an industry. The professional societies are key institutions in the documentation and transfer of technological knowledge.

10. ROLE OF THE ADMINISTRATOR AND THE ENTREPRENEUR

The administrator and manager play much more important roles in applied research than in basic research. The administrator of applied research must be more concerned with the substance of the work than is the administrator in basic science, where his job is primarily to assure a proper environment for creativity with relatively little concern for the strategy of the research itself, except as an equal participant. An exception, of course, lies in the management of big science, where extensive forward planning, and therefore highly sophisticated technical judgment, are required.

The fundamental problem in managing applied research is

to provide support that will attract attention and focused effort, but still leave sufficient scope and freedom to reconfigure goals and reformulate problems as the work advances. The good research director seldom manages in the sense of telling people what to do; rather he manages indirectly by showing interest in the lines of research that appear to him most promising and, by himself, persuading people of the desirability and attainability of his own goals. In this respect, there is no substitute for a demonstrated record of past success in enabling a manager to bring his own people along with him intellectually. An important part of the task of a research director is to match science to the nonscientific goals of the organization.

Behind many new developments in technology one finds a dedicated enthusiast, a singleminded individual with a clear and unswerving vision of the goal that he sees as possible and desirable. For rockets, it was Robert Goddard; for transistors, it was William Shockley; for jet engines, it was Frank Whittle; for nuclear-powered ships, it was Hyman Rickover; for hydrogen weapons, it was Edward Teller. In some cases such individuals are swimmers against the tide of respectable scientific opinion, but in others they have the intellectual support if not always the shared enthusiasm of their colleagues. Unfortunately, for every such dedicated individual who is right there are many who are wrong, or even incompetent. In retrospect we tend to recall only those who were right, while the mistakes and misguided enthusiasms or enthusiasts are buried in obscurity. Perhaps the best test in the long run is the ability of such a man to attract other men of high intellectual caliber to his technological banner. Frequently such an individual is not himself the technological innovator, but the individual who, concentrating on a single and simply expressed goal, succeeds in attracting creative people to his own objectives. In this sense, he is a technological entrepreneur.

Little is really known about the characteristics of such technical entrepreneurs, although their importance to technological innovation is evident from almost every case study. They are a distinctive breed that does not fit a mold. Often they are lacking in the orthodox educational qualifications, and frequently they may appear scientifically ignorant or nonrigorous. They tend to be eclectic in their interests and tastes, not constrained by the formal boundaries between disciplines.

On the other hand, it would be dangerous to assume that all technical innovation requires such an entrepreneur. In many situations groups of talented people come together to work on new technology without any visible "entrepreneurs" or intellectually dominant individuals. It would be equally dangerous to assume that enthusiasm is a guarantee of success. The technical entrepreneur is an intellectual gambler who plays for big stakes, whereas the scientist is usually a man who, by temperament, plays for somewhat smaller stakes, but often with surer progress.

INDEX